Ethical Literacies and Education for Sustainable Development

Olof Franck • Christina Osbeck
Editors

Ethical Literacies and Education for Sustainable Development

Young People, Subjectivity and Democratic Participation

palgrave
macmillan

Editors
Olof Franck
Department of Pedagogical
Curricular and Professional Studies
University of Gothenburg
Gothenburg, Sweden

Christina Osbeck
Department of Pedagogical
Curricular and Professional Studies
University of Gothenburg
Gothenburg, Sweden

ISBN 978-3-319-49009-0 ISBN 978-3-319-49010-6 (eBook)
DOI 10.1007/978-3-319-49010-6

Library of Congress Control Number: 2017937888

Printed on acid-free paper

This Palgrave Macmillan imprint is published by Springer Nature
The registered company is Springer International Publishing AG
The registered company address is: Gewerbestrasse 11, 6330 Cham, Switzerland

Preface

Four Tracks

This book highlights challenging perspectives on how to embrace the development of sustainability education, and focuses on ethical dimensions which transcend the four key cornerstones that are often mentioned with reference to such an education: the ecological, the economic, the social and the cultural.

One track within the chapters illuminates issues relating to the classical questions: what, when, how, for whom and by whom is sustainability education to be embraced with regard to ethical perspectives and ethical challenges in our time? Sustainability education is always, like all kinds of education, situated within certain ecological, social, cultural and economic contexts, which can be viewed and interpreted from various historical and pedagogical angles. There is no sustainability education as such; there is always room for interpretation, for discussion and for critical examination. This holds not least when it comes to questions about how the concept of sustainability may be understood across time and contexts.

A second track highlights a supposed tension between analytical and normative interpretations of the concept of ethical literacy. What is the aim of education which highlights the ethical dimensions of, for example, ecological or social sustainability? Is it to foster pupils into more or less well-defined patterns of moral behaviour and moral action, or is it to offer opportunities to carry out creative and independent analyses of diverse conceptions of good and evil, right and wrong? Is the aim for such education to perform the task of relating the concept of sustainability to more or less traditional ways of approaching the question about what kinds of ability, skill or literacy pupils should be given opportunities to develop within an education where norms and values are in focus?

A third track explores the—ethical—question about what apprehension of young people´s integrity and freedom are represented in various positions with regard to the tension mentioned. Will the focus be on "children's right to share and engage in sustainability development in various areas"? Will it be on an inspirational, but perhaps indoctrinatory, level where the practising of relevant action competence is highlighted as the main aim for education which contributes to the development of concepts of sustainability, or will the focus instead be on the striving for a presumptively impartial and objective education where the teaching aims to give pupils opportunities to discuss relevant facts, approaches, investigations and theories in order to develop personal standpoints and positions, whatever these will be?

A fourth track centres on discussions about the concept of sustainability education itself, and not least with regard to the commonly used formulation "education for sustainable education". In particular, critical analyses of what may be conceived of as a "postmodern concept of sustainable education" will be in focus. Such a concept promises visions of a united dream of future harmony, justice and even love; a dream which is supposed to be dreamt by all and anyone, in spite of the fact that mighty and powerful interests will clash and collide wherever the concept of sustainability is used to catch sight of what needs to be done in order to make the world a better place to live in for humans as well as non-humans.

These four tracks together represent an approach which makes room for theoretical as well as practical, and critical as well as constructive, presentations and analyses that could be of interest and of use to anyone engaged in sustainability education or research in this field. The ethical dimensions transcending the four key cornerstones—the ecological, the economic, the social and the cultural—and highlighted in the book satisfy a need and contribute to filling a gap with reference to educational as well as research-based analyses.

The Chapters

In Chap. 1, Olof Franck examines what is described as the challenge of developing democratic education for sustainability, where the aim is not that the pupils should be fostered into taking specific moral position, but rather that they should become aware of the right to deliberately choose ethical actions and strategies as moral and social subjects. At issue are various concepts of ethics education elaborated with reference to sustainability education. The often observed tension between "analytical" and "normative" approaches to education about ethical issues is claimed to be neutralizable if the focus is on the prerequisites for the pupils' becoming engaged moral subjects rather than on expected "results" in terms of "moral positioning" and "moral action". The concept of subjectivity is discussed with regard to the philosophical-pedagogical approach developed by Gert Biesta. Other inspirations for the development of the argument are Jacques Rancière and John Wall.

Chapter 2 by Marie Grice explores relational perspectives of transdisciplinarity and education from a theoretical angle to raise epistemological and ethical issues regarding ESD. *Philosophizing with* is then used as an analytical tool in the epistemic exploration from concept to educational context where systems thinking provides common ground for ethics and ESD competence. Through critical thinking and moral literacy, teachers are claimed to understand the role of ethics in the co-creation of knowledge. Teachers and students learn through a relational process involving epistemic, ethical and practical boundary crossing in which the educative moment, *le moment*, might reveal itself.

In Annika Lilja's Chap. 3, the focus is on the school's task to foster pupils' ability to develop knowledge, and also values such as equality, solidarity, human rights and democracy. It accentuates the difficulties of teaching ethics in school in ways that influence the pupils' resources for taking action. Teachers testify that during lessons pupils reason wisely about how to act in a moral way and how to choose between good and bad, right and wrong. Then in breaks the pupils go out in the schoolyard and do just the opposite. It is as if the pupils do not see the connection between what they discuss during lessons and how to behave in practice. The chapter reports on how ten pupils talk about how they learn moral competence in school and how they value education in school.

Chapter 4 by Christina Osbeck investigates what kinds of ethical discourse of sustainability are offered to tweens in Sweden through fiction reading. An empirical analysis of four books is presented. Ethical literacy is here understood in relation to sociocultural perspectives of moral development, and it is perceived as being connected to the repertoires of discourses that the individual has access to through the discursive practices where they are active, but also as related to the specific context of a situation—that is, what discourses can be privileged and used in this context. Fiction reading may, it is argued, be a way to expand one's repertoires of discourses beyond those which one appropriates through the discursive practices where one lives one's everyday life. The special opportunities that fiction reading offers when it comes to expanding one's available discourses are elaborated with reference to Mark Tappan's socio-cultural approach and Martha Nussbaum's philosophical and virtue ethical theories.

In Chap. 5, Anna Lyngfelt investigates what opportunities children have to act as moral subjects within the school context, while working with fiction at school. Theoretically, Edmund Husserl, Maurice Merleau-Ponty and George Herbert Mead are referred to, since they, by contributing to phenomenological theories of "acts", seek to explain in what ways social agents constitute social reality through language (and other symbolic social signs). The idea that discourse creates linguistic structures to construct the self is essential, as well as the concept of identity as unstable and non-coherent; identity is regarded a performative accomplishment, where the act that one "does" (performs) does not start with classroom discourse. To be able to discuss the opportunity to act as moral subjects, Lyngfelt discusses fiction and focuses on social issues from a variety of positions. The texts analysed are all extracts from textbooks used in Swedish compulsory schooling, discussing moral standpoints stressed by the curriculum (i.e. equality in terms of gender, ethnicity and social background). Since the extracts are presented together with exercises in the textbooks, the result of the analyses of text extracts (including exercises) are compared with analyses of the full-length texts. In this way the opportunities for students to act as social subjects are discussed, as well as the constraints of the classroom discourse.

Chapter 6 by Margaretha Häggström examines life-world perspectives on art-based environmental education (AEE) and sustainability. In focus in her discussion is the view of so-called "plant blindness" and how school pupils might connect with natural environments through AEE. The chapter starts out from a phenomenological perspective with a focus on the concept of the life-world and how to understand plant blindness in relation to pupils. The main purpose is to contribute to the debate about plant blindness by discussing pupils' life-worlds concerning AEE. AEE is here seen as a didactic tool to embrace bodily knowledge and to meet the demands of sustainable education, according to the United Nations Educational, Scientific and Cultural Organization's (UNESCO's) policy documents and Swedish curricula. One conclusion from the discussion is that nature and sustainability need to be experienced bodily repeatedly to be a part of pupils' life-worlds.

In Chap. 7, Kassahun Weldemariam traces the discourse within early childhood education for sustainability (ECEfS) from historical trend, policy, curriculum, pedagogy and research perspectives. Intrigued by contemporary post-human thinking, he poses fundamental questions about the ontological, epistemological and ethical starting points which call for a rethinking of sustainability discourse at large and ECEfS in particular. Drawing on the post-human perspective, the chapter identifies and elucidates one of the deep-rooted and inherent problems within environmental sustainability discourse—its human-centric characteristics—and indicates the need to recognize the more-than human and the non-human world. In so doing it challenges the existing dominant human-centric discourse and calls for the ECEfS field to consider and adapt to the ontological, epistemological and ethical shifts that the post-human turn has introduced.

Dawn Sanders' Chap. 8 examines the window and its environs as a point of departure for children's sense of agency and sense of place between home and the wider world. The discussion is framed by three children's books from Sweden, Australia and the UK: *Linnea's Windowsill Garden* (Björk 1978), *Window* (Baker 1991) and *The Flower* (Light 2006). Two of the books focus on growing plants; Björk makes border crossings between fiction and non-fiction in an intergenerational story of a girl and her grandfather growing a windowsill garden in their apartment, while Light offers a dystopian view of a grey monochrome world in which growing flowers from seed is seen as a transgressive act. Baker presents the view through the window as "a metaphor for the changes happening in the world". The books were chosen because they represent two ethical concerns for contemporary childhood: sense of agency and sense of place. Sanders explores how each book can be interpreted as an ethical tool in a range of learning contexts, both in and beyond the classroom.

In Chap. 9, Eva Nyberg investigates how aesthetical and ethical dimensions seem to be present in sensory experiences that constitute the basis of parts of science education. With reference to her previous research and, not least, to her long experience as a teacher in science education, she develops a reconstructive analysis of how the basis mentioned could be interpreted and elaborated. Starting with the indications from her earlier studies that taking care of, and observing, living things in the classroom can elicit emotional values and have a positive impact on pupils' interest in organisms that they previously considered uninteresting, Nyberg discusses how both aesthetic and affective experiences, through personal encounters, observations and guided explorations, seem to play a role in children's perception of plants. She presents and explores two case studies, one with 11-year-old children studying blueberry plants during an outdoor excursion; the other with a group of student primary teachers growing and observation pea plants. The data consist of diaries, photos and video observations.

Research and Educational Practice

The chapters in this book call attention to research-based examinations as well as pedagogical approaches with reference to educational practice in, for example, the use of children's books in sustainability teaching, and ethical and aesthetical dimensions of environmental education. One area rarely treated with reference to educational contexts—namely, post-human perspectives within early childhood education—is carefully discussed and analysed in one chapter. In another, phenomenology is originally used to bring AEE and sustainability into focus, with special regard to a life-world perspective.

A key benefit of this volume is that ethical dimensions of sustainability education are raised, analysed and discussed with respect to various relevant challenges to such education. By focusing on the concept of ethical literacy, which is central to issues regarding sustainability and sustainability education in practice, and with reference to relevant research areas, the chapters offer valuable contributions to the ongoing discussion about ethics, education and sustainability.

This book aims to present fresh and creative perspectives on how ethics and issues regarding ethical literacy can be elaborated with respect to democratic sustainability education. The intended audience is researchers and practitioners (teachers) in the fields of ethics and sustainability education, but the project should also be of interest to students participating in, for example, teacher education.

Department of Pedagogical
Curricular and Professional Studies
University of Gothenburg
Gothenburg, Sweden

Olof Franck
Christina Osbeck

Acknowledgements

The editors would like to thank the Department of Pedagogical, Curricular and Professional Studies at the University of Gothenburg for its kind support during the work on this book, not least by arranging writing retreats where the staff were invited to participate by writing articles in peace. Many of the contributors to this volume have joined in these week-long sessions and developed their text while at the same time having the opportunity to discuss themes and research questions with their colleagues.

Contents

Abbreviations

ACECQA	Australian Children's Education and Care Quality Authority
AEE	Art-based environmental education
DE	Descriptive facts about ethics (approach)
ECE	Early childhood education
ECEfS	Early childhood education for sustainability
ESD	Education for sustainable development
MF	Moral fostering (approach)
OMEP	World Organization for Early Childhood Education (Organisation Mondiale Pour L'Éducation Préscolaire)
PE	Philosophical ethics
RE	Religious education
UN	United Nations
UNEP	United Nations Environment Programme
UNESCO	United Nations Educational, Scientific and Cultural Organization

List of Figures

List of Tables

Biographical Notes

Olof Franck is Associate Professor of Philosophy of Religion and Senior Lecturer in Subject Matter Education in Social Studies at the Department of Pedagogical, Curricular and Professional Studies, University of Gothenburg, Sweden, where he also is Assistant Head of research. He has been engaged as a subject expert by Skolverket in the school reforms Lgr11 and Gy11, and in the national IT strategies project with a focus on digital competence in Swedish policy documents for upper secondary school. He is the author of a range of articles and books relating to the philosophy of religion, religious education, ethics education, and issues regarding multicultural and gender perspectives on education. His main research focuses on various issues within ethics education, religious education and education for sustainable development (ESD). He is engaged in a research project financed by a grant from the Swedish Research Council for 2015–2017: What may be learnt in ethics? Varieties of conceptions of ethical competence to be taught in compulsory school (Dnr 2014–2030).

Margaretha Häggström is a doctoral student in pedagogy in the Faculty of Education, University of Gothenburg, Sweden. She began her PhD work in 2015 after four years of teaching at the teacher education programme at the same university, six years of teaching in high school and twelve years in preschool. Her subjects are visual arts and Swedish (language and literature). She has a multimodal perspective on teaching and learning, and her research focuses on aesthetical values in relation to students' experiences of being surrounded by natural settings. She is also editing a book that addresses multimodality and visual knowledge as teaching and learning tools in all school subjects.

Annika Lilja has a PhD in pedagogical work. She has many years of experience as a teacher in Swedish compulsory school. She has also been working with the national tests in religious education for some years. Her research interests include pedagogical relations, and issues to do with values education and ethics education. Between 2015 and 2017 She is engaged in a research project financed by a grant from the Swedish Research Council for 2015–2017: What may be learnt in ethics? Varieties of conceptions of ethical competence to be taught in compulsory school (Dnr 2014–2030). She is also Director of Studies in the Department of Pedagogical, Curricular and Professional Studies, University of Gothenburg, Sweden.

Anna Lyngfelt is Associate Professor in the Department of Pedagogical, Curricular and Professional Studies, University of Gothenburg, Sweden. She has 20 years of experience in school education research, including research on teaching and learning in primary and secondary schools. She also has experience in school governance and innovative school practices, and pedagogies and teaching methods. Her research projects for 2012–2015, *Läsa mellan raderna* [Reading between the lines] and *Digitala arenor i läs- och skrivpraktiker i grundskolans tidigare år* [Digital arenas in literacy practices in early primary school] were both intervention studies, including teachers. She has experience of qualitative research methods and the combination of quantitative and qualitative research methods, especially from her work with reading assessment and literacy development. Originally being a teacher at secondary school, she has a MA in language education.

Eva Nyberg is a PhD and Senior Lecturer in the Department of Pedagogical, Curricular and Professional Studies, University of Gothenburg, Sweden. Her teaching and research concern science education, especially biology, in primary and secondary school, and in higher education, especially aesthetic and affective dimensions in teaching and learning science. She is also interested in environmental and sustainability education throughout the educational system, as well as outdoor education. Through her thesis work she developed an expertise and interest in formative assessment, especially in the context of teachers' professional development. She is engaged in a research project focused on learning about plants: Beyond plant blindness: Seeing the importance of plants for a sustainable world, funded by the Swedish Research Council (Dnr 2014–2013). She has extensive experience of teaching in upper secondary school, in teacher education and in in-service training of teachers.

Christina Osbeck is a PhD, Associate Professor in Religious Studies and Senior Lecturer in Social Studies Education, as well as Deputy Head in the Department of Pedagogical, Curricular and Professional Studies, Gothenburg University, Sweden. Her research is focused primarily on religious education and values education, which may both be said to include issues relating to ethics education. Ethics education is also a central aspect of the department's research in education for sustainable development. Her main research at present concerns the project What may be learnt in ethics? Varieties of conceptions of ethical competence to be taught in compulsory school (Dnr 2014–2030), of which she is the principle investigator, funded by the Swedish Research Council,. She has been working in teacher education since 1993 and her dissertation, delivered in 2006, focuses on pupils' life interpretations, and how bullying is working as a shaping and homogenizing tool of life interpretations.

Dawn Sanders is Associate Professor in the Department of Pedagogical, Curricular and Professional Studies, University of Gothenburg, Sweden, and a Fellow of the Linnean Society of London. Her doctoral study at Sussex University in 2004 examined the educational role of botanic gardens, a study inspired by her ten years as an educator at the Chelsea Physic Garden, London. She trained as an artist before completing her botanical education and draws on both in her academic work. She is a scientific associate of the Natural History Museum, London. Her recent publications include *Darwin-Inspired Learning* (Sense, 2015), of which she is both a co-editor, along with Prof. Michael Reiss and Dr Carolyn Boulter, and author/co-author of four chapters. Her chapter in this anthology draws on her teaching work with early years educators at Gothenburg University and her current research work, specifically Beyond plant blindness: Seeing the importance of plants for a sustainable world, funded by the Swedish Research Council (Dnr 2014–2013), a cross-disciplinary project for which she is principal investigator.

Arjen Wals is Professor of Transformative Learning for Socio-Ecological Sustainability at Wageningen University, the Netherlands, andGuest Professor in Education for Sustainable Development at Gothenburg University, Sweden . He also holds the UNESCO Chair of Social Learning and Sustainable Development and is an Adjunct Faculty member at Cornell University,USA. His teaching and research focus on designing learning processes and learning spaces that enable people to contribute meaningfully to sustainability. A central question in his work is how to create conditions that support (new) forms of learning which take full advantage of the diversity, creativity and resourcefulness that is all around us, but so far remain largely untapped in our search for a world that is more sustainable than the one currently in prospect. In 2014 he was the lead author of an article published in *Science* on the role of citizen science in bridging science education, environmental education and sustainability. He is editor or co-editor of a number of popular books, including *Higher education and the challenge of sustainability*" (2004), *Creating Sustainable Environments in our Schools* (2006), *Social Learning towards a Sustainable World* (2007), *Learning for Sustainability in Times of Accelerating Change* (2012), and of the *International Handbook on Environmental Education Research* (2013). He has (co)authored more than 250 publication of which 60 are in international peer-reviewed journals. He writes a regular research blog that signals developments in the emerging field of sustainability education (www.transformativelearning.nl).

Kassahun Weldemariam is a PhD candidate in the Department of Pedagogical, Curricular and Professional Studies, University of Gothenburg, Sweden. He holds an international master's in early childhood education and care and a master's degree in teaching english as a foreign language. He worked as a teacher trainer for four years at Dilla University, Ethiopia, and as a preschool pedagogical leader for three years in Stockholm, Sweden. The working title for his PhD is "Children's relations/entanglement with materials as a stepping stone for environmental sustainability: New-materialism perspective". His work is based on a critical view of the human-centric and environmentalist approach to sustainability in early childhood and it attempts to embrace the non-human actors in the endeavour towards sustainability. His earlier publications include "Cautionary tales on interrupting children's play" (2014, *Childhood Education*, 90(4), 265–271) and "Professionalism in early childhood education and care in Ethiopia: What are we talking about?" (2013, *Childhood Education*, 89(3), 152–158).

Highlighting Ethics, Subjectivity and Democratic Participation in Sustainability Education: Challenges and Contributions

Olof Franck

Introduction: Challenges of Democratic Education

In this chapter the challenge of developing democratic education for sustainability, where the aim is not that the students will be fostered into taking specific moral positions but rather that they will become aware of the right to deliberately choose ethical actions and strategies as moral and social subjects, is highlighted.

Democratic education constitutes an arena where it is of great importance to continuously discuss its fundamentals and its prerequisites in the light of threats that might challenge the idea of performing education which satisfies basic democratic standards. Not least within ethics education are such threats significant.

One challenge is shaped on the fact that democratic relations between grown-up teachers and young students might be threatened if and when the former are treated as the omniscient and active, and the latter as the

---

O. Franck (✉)
Department of Pedagogical, Curricular and Professional Studies,
University of Gothenburg, Gothenburg, Sweden

O. Franck, C. Osbeck (eds.), *Ethical Literacies and Education for Sustainable Development*, DOI 10.1007/978-3-319-49010-6_1

passive receivers of more or less definite ethical knowledge. Another challenge stems from the risk that undemocratic—or pseudodemocratic—relations between teachers and students take the form of indoctrination in the sense that students are taught how to act in moral matters without having an opportunity to question and criticize the reasons and intentions behind such prescriptive teaching. A third challenge is that indoctrinatory, undemocratic ethics education might become focused on policy issues where all that seems important is to get the students to reach measurable outcomes by handling specific decision scenarios, with no room for philosophical reflection and creativity (Hartner 2015; Franck 2017).

These challenges are more or less visible also with regard to democratic sustainability education, which is not surprising because ethics is counted as one of the main dimensions of sustainability, and one that relates in various ways to the strands often mentioned: the environmental, the social, the cultural and the economic (UNEP 2015). In this chapter I shall discuss how these challenges can be interpreted, and I will also say something about what kinds of strategy might be relevant when trying to address them, considering that sustainability education often seems to be conceived as a forum for the transmission of values. The aim is to elaborate on a concept of democratic sustainability education where the development of ethical skills is in focus with regard to all participating actors, and where such a development is shown to constitute a firm but flexible basis for significant teaching about sustainability.

Classifying Ethics Education

Before outlining a possible classification of various strands within sustainability education, I shall, as a starting point, highlight a characterization of the aims and structure of ethics education. This characterization has its roots in a distinction of religious education (RE), presented by Michael Grimmitt and Garth Read (1975). Starting the classificatory outline in this way is relevant because ethics has a natural place within both sustainability education and RE.

Grimmitt and Read (1975) make a distinction between two dimensions of learning in RE: learning *about* and learning *from* religion. Learning about religion refers to learning within phenomenological teaching about world religions, teaching which is not thought to lack space for critical discussion and analysis but which makes room for an "emphatic" study of religious thought and religious traditions (ibid.). Learning from religion explores an existential dimension which is built around an apprehension according to which theological ideas and dogmas are not to be seen as the primarily important content within RE. Rather, a wider context mirroring various strands of human faith and beliefs, including ethical apprehensions, moral practice and existential issues regarding meaning and purpose in life, is what should be at the centre of RE. According to Grimmitt, both dimensions ought to be involved in RE (1987).

A prerequisite for such double-dimensioned education is, to quote one commentator, an emphasis not of "knowledge of religion per se ... but the way in which a religious believer perceives the world and how these insights can inform how the learner sees the world" (Teece 2010). What should be in focus is the aim of opening up for identification processes where learners, reflecting on fundamental existential and ethical issues, might become inspired by studying religious believers' ways of handling such issues, recognizing some of these ways, perhaps, while not being able to understand or feel comfortable about others. Moreover, here Grimmitt stresses that it ought to be a mutual interpretive process—in the sense that learners within secular RE might be influenced both by knowledge about how universal existential and ethical questions are dealt with by religious believers and by their own life experience—that puts new interpretive dimensions and perspectives in focus in RE. This is, in fact, what is meant by "emphatic" education about religions, through which learners can develop the ability to identify with life experiences of religious believers, and their existential and ethical interpretations of the same. Students will also be able to recognize such interpretations as less alien and less far-reaching than otherwise (Grimmitt 1987).

It is possible to use Grimmitt's distinction, at least as a tentative frame of reference, as a means of classifying ethics education. Such education *about* ethics would mean that various issues and questions are highlighted and made the objects of reflection and analysis, but without any demands being raised that these examinations should pave the way for a more or less explicit moral positioning, putting forth personal standpoints and opinions. Education from ethics would, following Grimmitt's model, be performed with the aim that the participating students should be able to relate to more or less fundamental ethical issues by having the opportunity to identify with people engaged in moral struggles. They could then try to find ethical, or ethically relevant, tools to handle challenges of various kinds, and search for a basis with regard to which moral choices could be made in their lives.

Rationality and Prescriptive Dimensions of Ethics Education

However, the above classification of ethics education suffers from two shortcomings. First, it is based on the notion that people in their moral lives, more or less consistently, act like rational agents who weigh reasons for and against before, on supposedly objective grounds, they consider how to address the challenges they face. It may be true that we all, though certainly not always, strive to try to choose constructive ways to meet life's challenges and opportunities. However, such choices cannot be interpreted in purely rational terms: they are constituted of thought, reason, emotions, attitudes and values, and are therefore complex (Franck 2017).

Far too often it appears that schools' ethics education is based on a picture of what it means to be a moral agent. Students receive information about scenarios in which they should consider different options to deal with moral dilemmas—dilemmas that are often located quite far from their everyday lives. Their comments must be justified by an ethical theory or an ethical model, but such theoretical or "objective" considerations represent only one dimension of what it means to make moral decisions. Therefore this kind of teaching does not reflect people's everyday moral life (Skilbeck 2016).

The second shortcoming is that the classification in question does not provide sufficient tools to be able to capture the breadth of the perceived ethics teaching. Grimmitt mentions in connection with his categorization that besides learning *about* and *from* religion, there is a confessional teaching of "pure religion" (1987). As has been pointed out (Teece 2010), it can be formulated as education *in* religion.

As for non-denominational ethics education, it is essential to highlight teaching whose aim is that students should be notified of norms and values that they are expected to follow. In focus is an ethics instruction based on more or less explicit ethical axioms on which the teaching is performed. Here a need for a broader model of categorization is worth considering when it comes to classifying sustainability education.

Modelling Categories of Ethics Education

To build a classification of sustainability education on a broader model for categorization, that is to say one that includes an ethics instruction dimension, I shall use a proposal presented by Gardelli et al. (2014). This proposal of categorization of ethics education highlights three approaches that are supposed to capture different ways of understanding the teaching of ethics: (1) the descriptive facts about ethics (DE) approach; (2) the moral fostering (MF) approach; and (3) the philosophical ethics (PE) approach (p. 16).

Regarding "ethics in school", the DE approach is said to refer to a concept according to which education will be directed towards social facts about people's ethical behaviour and reasoning in morals. Students can, within such a kind of ethics education, examine how certain groups or individuals take a stand on different ethical issues and act in different social contexts. The aim of such an approach is "teaching (or helping the student learn) social, statistical, psychological, or sociological facts about moral issues" (p. 17).

The MF approach is said to refer to a range of interpretations and concepts. However, Gardelli et al. choose to use the label in a "rather narrow" sense, more precisely as the transmission of values in "some kind of rather straightforward" sense, focusing "moral fostering" (p. 18).

Finally the PE approach is said to highlight not empirical or sociological studies of people's moral attitudes or behaviour or the fostering of students to accept certain norms and values (p. 18), but normative skills, such as the formulation of arguments for and against moral positions, and the ability to make assessment judgements with reference to the evaluation of reasons (p. 18).

After having discussed these approaches, Gardelli et al. reach the conclusion that "ethics in school" should be built primarily around the PE approach, while also giving recognition to the possibility of making room for some moral fostering (p. 25) and that there might be other "strong arguments" to consider with regard to ethics education which could lead to "other conclusions" (p. 25). The DE approach is left out since it does not focus on is what important in school contexts—namely, "engaging in" ethical matters and "*doing* normative ethics" (p. 18).

I do not, in the present context, want to go further into the discussion presented by Gardelli et al. However, I mention their distinction to help in trying to categorize ethics education—and, as we shall now see, also in the classification of sustainability education where dimensions of ethical skills are highlighted.

Education from Sustainability

The label "sustainability education" is, in the present context, for pedagogical reasons, replacing "education for sustainable development (ESD). Leaving the discussion regarding possible challenges related to the concept of "development" behind (Knutsson 2014), one might ask what kind of aims and content would be significant for sustainability education? Is it education *in* sustainability? Would the aims and the content differ from education *about* and *from* sustainability? If so, in what sense?[1]

If we choose to talk about "education *in* sustainability", we shall, firstly, find ourselves forced to consider some challenges that are the same as those associated with ESD. Certain more or less well-defined values and norms are assumed to be important to transmit to the students involved, and then there appear to be opportunities to introduce activities that are interpreted to be consistent with those values and standards.

However, such a concept of sustainability education threatens other fundamental norms and values, especially those often characterized as key-stones in democratic communities: personal freedom and integrity, and the right of individuals to make free choices, to mention just two. Many writers have criticized ESD for including prescriptive methods and aiming to nurture what has been described as "eco-certified children" (Ideland and Malmberg 2015), and thus failing to take into account a student's right to be critical of the sustainability education they are forced to participate in and of the normative and value-based aims governing it (Jickling 1994; Dahlbeck 2014; Davies and Elliott 2014).

Second, education in sustainability is connected to another and no less acute challenge—namely, the one that the term sustainability is porous and allows for a plethora of interpretations. "Sustainable development" has been criticized on many grounds, not least for signifying a Western concept of developmental optimism (Hellberg and Knutsson 2016). Changing the concept from "sustainable development" to "sustainability" may help to shift slightly the emphasis in the education we are talking about, but previous objections remain. The "post-political" consensus often signalled in the policy documents and speeches is a chimera which conceals the strong conflicts of interest and claims to power that control the processes that operate under the concept of sustainability (Rist 2008; Knutsson 2014).

What kind of sustainability education could serve as a trustworthy alternative to a normative and prescriptive one? From the discussion of the categorization of ethics education, we remember the criticism that Gardelli et al. hinted at in "education about ethics": it is not about *engaging in* or *doing* normative ethics. In a sense, this course is a relevant objection. On the other hand, "education about ethics" may perhaps be perceived to be not that defensive or even irrelevant. (Grice & Franck 2014) As we saw earlier, Grimmitt does not mean that teaching about religion is the most important thing in schools' RE. He understands, though, the knowledge which was treated as a prerequisite for the students to be able to implement engaged participation in the education on existential and ethical matters. Education *about* religion is, one could say, to be apprehended as relevant in an instrumental sense, and maybe one can—even if education *about* ethics can be described as having at least partly other content—say that education about ethics could be apprehended in a parallel way.

Given the fact that sustainability education includes fundamental ethical dimensions, one could consider whether it is not also the case that education about sustainability, if it can be categorized reasonably clearly, could be attributed an instrumental role? But, one might ask, is it at all possible to give a both reliable and more or less universally accepted categorization of the purpose and content of education about sustainability that is needed to be able to speak of it as "an instrument" for another and perhaps more important kind of sustainability education?

Can one imagine that it is reasonable to consider education *about* sustainability, where students might have access to historical and current discussions about how environmental, social, economic and cultural issues have been discussed in debates and documents that are described as more or less fundamental points of reference within sustainability discourses? Could it be considered instrumental in relation to education *from* sustainability, where these discussions, often in combination with relevant statistical data concerning emissions, energy consumption, poverty, economic imbalances and so on, are treated critically and with a focus on ethical perspectives, as well as theories in science and social sciences? Would this be a strategy that might lay the ground for the development of sustainability education? What is looked for is a development which crosses between the pitfalls of either failing to satisfy democratic and ethical demands of respecting students' freedom and integrity, or nurturing a kind of distant study where participants refrain from engaging in sustainability issues. The latter alternative would hinder the participants from doing normative ethics in the sense that they get the opportunity to critically examine and discuss ethical questions with regard to environmental, social, economic and cultural dimensions of these issues.

It is relevant and important to investigate this strategy, not least because it offers the opportunity to involve students as independent subjects in the teaching and learning processes to be elaborated under the designation "sustainability education". Here there is, at least in principle, a space for building a foundation on which democratic values can govern the aims, the structure and the content of the education in question. Students can be partners in both the planning and the implementation of courses, lessons, excursions and exhibitions, and the issues that are to be prioritized with regard to this or that consideration, which may be matters for democratic dialogue and decision.

I should like to emphasize that I am now describing a *possible* scenario which *might* be realized. It is important to discuss the roads along which we want democratic, ethically well-founded sustainability education to be structured. Which roads are to be found between education *about* and education *in* sustainability, given that "about" and "in", if they are to be interpreted in a bold and exclusivist way, both seem to challenge reliable and democratic sustainability education? Students have the right to come to know about information that is relevant to the development of personal, substantiated standpoints and the motivation for action. They also have the right to critically examine other standpoints and other actions than those that they have made their own. Education *from* sustainability, where "from" is understood as signifying a structure where the starting point for teaching and learning is students' existential and moral experiences in relation to whatever environmental, social, economic and/or cultural issues are treated and highlighted, is a candidate for filling a need for a democratic and ethically well-founded educational strategy.

Sustainability Education and the Transmission of Values

This is not, of course, to say that this strategy is unproblematic. In the next section I shall discuss one important difficulty. Even if we find the criticism of the normative and prescriptive approach discussed above reasonable, we could take a further step and ask: What exactly does this criticism mean? We may approve of the rejection of—undemocratic and indoctrinatory—aims and moves where students are forced to accept certain norms and values without resistance, but does that mean that we would also like to hold on to a position where all kinds of value transmission in education are thought to be objectionable?

This is probably one of the most challenging issues to examine with regard to a democratic and ethically well-founded sustainability education. It touches on important ethical, philosophical and pedagogical questions related to not only sustainability education but also education in general. It points to what might seem to be a frustration among teachers when trying to develop democratic education at the same time as they feel responsible for transmitting certain values: is it possible to shape a reasonable consistency between students' right to freedom, integrity and personal decisions in existential and ethical matters, and the nurturing of certain values which they, within the school system, are expected to shoulder (Osbeck et al. 2015)?

Aims and Outcomes, Processes and Results

A first issue to penetrate here is one that focuses on what aims there are for structuring sustainability education of the kind mentioned. All education reflects certain values that shape the educational organization in terms of courses, lessons, exercises, tests and so on. If the guiding principle here first and foremost, explicitly or implicitly, is directed towards achieving specified objectives in terms of "results" and "outcomes", there will be a risk that the value and the importance of the teaching-learning process, where teachers and students cooperate in reflection, discussion and dialogue, is underestimated—and, worse, not prevalent at all. Such a concept of education where measurability is thought of as the solution to all kinds of challenge facing today's teachers and students does not lay a foundation for well-founded, reflective knowing. Such a knowing does not only reproduce dominating apprehensions of what "knowledge" and "values" are and should be, but also makes room for the criticism and realization of emancipatory visions and ideals (Franck 2017).

In the foregoing analysis a philosophically structured education *from* sustainability was tentatively sanctioned, at the same time as it was emphasized that the teaching-learning *in* sustainability, including values that frame an ethical basis for sustainability concepts, should be taken into account. It is fundamental that education *from* sustainability is governed not by focusing on specific outcomes or results but rather within a communicative process where the participants—teachers as well as students—are engaged in a dialogue regarding what aims should be highlighted, given present and historical conditions they find relevant to a constructive and reliable cooperation in sustainability issues, and how these aims are going to be applied within the education in question.

This is a guiding principle which harmonizes with a concept of a democratic education where *democracy* is structuring education rather than apprehended as the intended outcome. One representative of such a concept is Gert Biesta, who has emphasized the importance of education to open up arenas where participant students will have, and can see, opportunities to advocate and develop as knowing and acting subjects. It is through differences that humans are able to develop a subjectivity (Biesta 2003), and "diversity" may refer to age, gender, ethnicity, religion, political affiliation, sexuality, health and so on.

Being and *becoming* are to my mind fundamental concepts to elaborate on when trying to understand what is at stake here. Within democratic educational contexts, students are realizing that they are part of the community with the same rights and duties as all participants, and this means that they take part in the educational processes regarding structure as well as content. *Being* part of the societal community, they, like other participants, are expected to think and act as responsible agents, not in order to reach an aim definable in terms of "democracy" in the future but by practising democracy here and now (ibid.). Such practice will do something with the subject. They will act in a state of *becoming*, developing a subjectivity in relation to other agents (Franck 2016), The issue for democratic education is, as Biesta states, "not about how to 'create' or 'produce' democratic citizens, but about how to create opportunities for action, for being a subject, both in schools and other educational institutions, and in society as a whole" (Biesta 2003, p. 59).

Equality as a Prerequisite in the Relational Educational Process

According to the approach described above, democratic educational processes can be pictured as communicative processes where dialogue, understood perhaps in deliberative (Englund 2007) or agonistic (Mouffe 2005) terms, structures the relations between the participating subjects. This creates a need to highlight another issue with regard to education *from* sustainability—namely, that concerning teachers' and students' collaboration in educational processes.

In *The ignorant schoolmaster: Five lessons in intellectual emancipation*, the French philosopher Jacques (Rancière 1991) pictures the structure of a communicative educational process where teachers and students, within the framework of teaching, together are trying to grasp something that for them is unknown. When they thus collaborate and jointly try to understand, according to Rancière, the equality that exists between them is manifested. They go into the knowledge process as equals, even if they come from different places with different experiences and thoughts.

Rancière criticizes a concept according to which teachers are seeking to lower themselves to the learner's level in order to impart knowledge to them that will grow step by step according to a definite course of development where the goal, the "outcome" or the "result" is that the student should achieve a body of knowledge and quality, measuring themselves against those of the teacher. Equal education is about liberation (cf. Säfström 2015).

Rancière has something important to contribute when we consider how teachers and students can interact in a democratic manner in a teaching-learning process regarding sustainability issues. If they go into this teaching as equals in that none of them makes a claim to have exclusive access to the knowledge that the others lack, they may shape an epistemological and existential project where curiosity and criticism together pave the way for the realization of emancipatory aims and visions. They can approach sustainability issues without either producing or reproducing established hegemonic frameworks where one of them, "the inferior" student, is expected to strive to approach the other, "the superior" teacher, and where the power to judge form, content and value of what is supposed to be worth knowing is reserved for the latter of the two (Franck 2017).

Children as Knowledgeable Equals

A third issue to be highlighted with regard to the structuring of education *from* sustainability concerns what might be demanded by a democratic educational system regarding educational relations between teachers and young students. Rancière seems primarily to be discussing questions of democracy with reference to examples from academic contexts. In this chapter, however, I am interested in an analysis of what democratic and ethically relevant sustainability education for children might mean. Are children treated as equals within the educational processes going on in such education? John Wall, an American advocator of "a childist approach", has argued that children's experiences and interpretations of life are too often set aside. Ethics and morals are perceived as disciplines that reflect adult life, relationships and challenges, and that must therefore be handled with the adult's proposed solutions. According to Wall, children are considered more or less consistently to be moral individuals according to the standards formulated and authorized on the basis of images of what adults find to be worth problematizing and their way of discussing issues of right and wrong, good and evil, and so on (2010).

Wall stresses that what needs to be done in a time when children are denied the right to act as morally full subjects is to try to improve their rights, protection and freedom (p. 2). Moreover, it is required that the whole ethos of the children and their relationship to adults must be reconstructed in a way that does not give unlimited focus on difference. Children's experiences of meaning, value, challenges, difficulties, setbacks, opportunities, trust, hope, security and so forth must constitute a basis for a general reflection of what life as a human being can mean. This makes the child's life interpretations not only interesting as objects of reflection, analysis or perhaps research; rather, the interpretation of the life of the child formulates help to influence how people—children as well as adults—perceive life, what is important and valuable, and what one ought to do to contribute to the development of good relations and a good society (p. 3–4).

Being and Becoming "Sustainabilists"

Wall's "childist approach" can be developed with reference to research done in preschool contexts, where young children interact with each other and with teachers, shaping teaching-learning processes that reach into epistemological and pedagogical, as well as ontological and existential, fields. The recognition that "playing" and "learning" cannot be the object of any exclusive distinction seems to be generally, or almost generally, established (Coates and Coates 2006).

Research has been done on preschool children's ability to learn mathematics within the complex discourses in which they and their teachers participate. One result from a few of those studies is that children experience when they are positioning themselves as "mathematicians", and such a positioning is encouraged and confirmed by the teachers—that is, that they *are knowledgeable* in mathematics (Lembrér and Meaney 2015, p. 6f.). However, this does not mean that they look on themselves as experts. On the contrary: by having self-confidence they can see both that they have considerable knowledge about a lot of important issues in the field of mathematics and that, by discovering where this knowledge has to be deepened and broadened in order for them to handle and solve more complex mathematical problems, there is still a lot to learn (p. 10). They are having the experience of both *being* and *becoming* mathematicians.

They are neither ignorant nor experts. They are participating in a teaching-learning process in which they are given freedom and responsibility to study relevant disciplinary issues, at the same time as they are developing their skills and competence in the field.

Perhaps it would be reasonable to assume that something like this is also going on in—democratic—teaching-learning processes within education *from* sustainability? If so, young children are to be conceived as *being* and *becoming* "sustainabilists" in the sense that they have the capability of discerning, reflecting on and assessing issues to be highlighted, interpreted, discussed, and critically and constructively penetrated within sustainability education. Not least, this may hold for the discernment of ethical dimensions: there is preschool research that supports the fact that very young children express and explore moral practice, in the sense that they show empathy and engagement in other children who feel sad and lonely, or who are in need of help (Johansson 2001).

Discussion

It has to be emphasized that the concept of education *from* sustainability, as it has been elaborated above, does not refer to a definite and fixed content, or to a methodological strategy which is planned and formulated in detail. The content in such education is to be negotiated in democratic, communicative teaching-learning processes. The same holds for the choice of educational methods. What is clear, however, is that education *from* sustainability will be influenced by certain general aims. These aims support the development of dynamic sustainability education where not only the concept of sustainability but also educational methods, strategies and approaches related to this concept are critically examined. Teachers and students are expected to discuss and criticize various concepts of sustainability, and also different approaches to establishing and developing pedagogies to reliably explore such concepts.

As mentioned at the beginning of this chapter, the concept of sustainability is porous. Any post-political stance according to which this concept refers, a content which may be the object of general agreement, will, as has been stated, obscure conflicts between powerful interests of different kinds, and claims of influence or even hegemony, which have to be analysed and critically investigated. This is one fundamental area to highlight within a substantial education *from* sustainability.

Another fundamental area within such education is the values which are thought to be related to the concept of sustainability. Not the least issues focusing what ethical norms and what moral practices that may be important to sanction and to follow in communities striving for sustainability, have to be critically and constructively examined within a democratic, communicative education *from* sustainability.

In this chapter, Biesta's approach regarding democracy and subjectivity in education as well as Rancière's interpretation of teachers and students as equal participants in educational processes have been referred to. Furthermore, Wall's emphasis of the need to include children's concepts of life, not in order to interpret or even judge them in terms of grown-ups' apprehensions but as important and valuable in themselves, has been discussed. We have also seen that there is interesting research regarding children's ways of positioning themselves as both *being* and *becoming* knowledgeable in teaching-learning contexts to study and to develop with reference to sustainability education.

Do these four references provide a foundation for forming a pattern which could serve as a base for a concept of education *from* sustainability, in which the challenge regarding how to handle the question regarding a need for the transmission of values may be taken care of? I think that they do, in a tentative way.

First, if the concept of children as independent subjects with the capability of moral discernment is accepted, there seems to be no democratically satisfying reason for not involving them as equals within a relational education *from* sustainability.

Second, the research on preschool children's positioning of themselves as both "being" and "becoming" mathematicians invites more comprehensive research if this self-characterization, including the fact that they see themselves as knowing things in the field while with regard to other relevant issues they have to study and examine relevant issues more, is possible to identify in the field of sustainability, including ethical dimensions.

This may be related to what Biesta has pointed out—namely, that teachers' activities within education may not be optimal if they exclusively take the form of refraining from giving hints, clues and suggestions as to how the subjects treated could be interpreted and developed, as long as the students have the opportunity to respond to and take a stand in relation to these (Biesta 2013).

Within education *from* sustainability, children and teachers will act together in communicative and democratic knowledge processes, and both are expected to take responsibility for contributing to creativity and development. More precisely, both may transcend formal and dualistic concepts of teaching-learning processes by sharing their experiences, knowledge and apprehension with each other.

It is within such a process that the "transmission" of values is to take place, not only in one direction but between the participants taking responsibility for a critical and constructive exchange of arguments, aims and visions. This is a process which presents not only challenges but also possibilities. First and foremost, it is important that it is continuously made the object of analysis, discussion and research. Education *from* sustainability will, from time to time and from place to place, be in need of being rethought, reinterpreted and reconstructed. This is neither a theoretical nor a practical defect; it is a demand which may militate against unsound conservatism, paving the way for undemocratic, unfair relations between grown-ups and children, between teachers and students.

Note

1. Compare this with the arguments presented in Kassahun Weldemariam's contribution to this volume (Chap. 7).

References

Biesta, G. (2003). Demokrati – ett problem för utbildning eller ett utbildningsproblem? [Democracy – A problem for education or an educational problem] Swe transl: E. Hultin. *Utbildning & Demokrati, 12*(1), 59–80. http://www.gu.se/infoglueCalendar/digitalAssets/1776442788_BifogadFil_Biesta.pdf

Biesta, G. (2013). *The beautiful risk of education*. Boulder/London: Paradigm Publishers.

Coates, E., & Coates, A. (2006). Young children talking and drawing. *International Journal of Early Years Education, 14*(3), 221–241. http://dx.doi.org/10.1080/09669760600879961

Dahlbeck, J. (2014). Hope and fear in education for sustainable development. *Critical Studies in Education, 55*(2), 154–169. http://dx.doi.org/10.180/17508487.2013.839460

Davis, J., & Elliott, S. (2014). An orientation to early childhood education for sustainability and research – Framing the text. In J. Davis & S. Elliott (Eds.), *Research in early childhood education for sustainability. International perspectives and provocations* (pp. 1–17). London/New York: Routledge.

Englund, T. (2007). *Utbildning som kommunikation: deliberativa samtal som möjlighet* [Education as communication: Deliberative communication as a possibility]. Göteborg: Daidalos.
Franck, O. (2017). Varieties of conceptions of ethical competence and the search for strategies for assessment in ethics education: A critical analysis. In O. Franck (Ed.), *Assessment in ethics education – A case of national tests in religious education.* Springer 2017, In Press.
Franck, O. (2016). Att finna vägar till att forma det goda: dygdetik som existentiell-moralisk metodologi [Finding ways of forming the good: Virtue ethics as existential-moral methodology. In O. Franck, C. Osbeck, & K. von Brömssen (Eds.), *Religioner, livsåskådningar och etik* [Religions, life-views and ethics]. Gleerups 2016, In Press.
Gardelli, V., Alerby, E., & Persson, A. (2014). Why philosophical ethics in school: Implications for education in technology and in general. *Ethics and Education, 9*(1), 16–28. http://dx.doi.org/10.1080/17449642.2014.890277
Grice, M., & Franck, O. (2014). A phronesian strategy to the education for sustainable development in Swedish school curricula. *Journal of Education for Sustainable Development, 8*(1), 29–42. doi:10.1177/0973408214529856.
Grimmitt, M. (1987). *Religious education and human development: The relationship between studying religions and personal, social and moral education.* Great Wakering/Essex: McCrimmons.
Grimmitt, M., & Read, G. (1975). *Teaching Christianity in RE.* Great Wakering/Essex: Mayhew.
Hartner, D. F. (2015). Should ethics courses be more practical? *Teaching Ethics, 15*(2), 349–368. http://philpapers.org/rec/HARSEC
Hellberg, S., & Knutsson, B. (2016). Sustaining the life-chance divide? Education for sustainable development and the global biopolitical regime. *Critical Studies in Education.* http://www.tandfonline.com/doi/full/10.1080/17508487.2016.1176064
Ideland, M., & Malmberg, C. (2015). Governing "eco-certified children" through pastoral power: Critical perspectives on education for sustainable development. *Environmental Education Research, 21*(2), 173–182. doi:10.1080/13504622.2013.879696.
Jickling, B. (1994). Why I don't want my children to be educated for sustainable development:Sustainable belief. *The Trumpeter. Journal of Ecosophy.* http://trumpeter.athabascau.ca/index.php/trumpet/article/view/325/497
Johansson, E. (2001). *Små barns etik* [Young children's morality]. Stockholm: Liber.
Knutsson, B. (2014). Utbildning för hållbar utveckling? Postpolitiska illusionsnummer och didaktiska alternativ [Education for sustainable development? Post-political illusory performances and didactical alternatives]. In O. Franck (Ed.), *Motbok – kritiska perspektiv på styrdokument, lärarutbildning och skola* [A book of critique. Critical perspectives of policy documents, teacher education and school] (pp. 177–195). Lund: Studentlitteratur.

Lembrér, D., & Meaney, T. (2015). Being and becoming as socialisation in a mathematical activity in preschool. *Tidsskrift For Nordisk Barnehageforskning/Nordic Early Childhood Education Research Journal, 11*(5), 1–14. Published 3rd of October 2015.

Mouffe, C. (2005). *On the political.* London/New York: Routledge.

Osbeck, C., Franck, O., Lilja, A., & Lindskog, A. (2015). Challenges of assessment in ethics – Teachers' reflections when assessing national tests. *EDUCARE – Vetenskapliga skrifter, 2015*(2), 19–47.

Rancière, J. (1991). The Ignorant Schoolmaster: Five Lessons in Intellectual Emancipation 1st Edition, Stanford university Press.

Rist, G. (2008). *The history of development: From Western origins to global faith* (3rd ed.). London: Zed Books.

Säfström, C. A. (2015). *Jämlikhetens pedagogik* [The pedagogy of equality]. Malmö: Gleerups.

Skilbeck, A. (2016). The personal and impersonal in moral education. In D. Lewin, A. Guilherme, & M. White (Eds.), *New perspectives in philosophy of education ethics, politics and religion* (pp. 59–76). London/New York: Bloomsbury.

Teece, G. (2010). Is it learning about and from religions, religion or religious education? And is it any wonder some teachers don't get it? *British Journal of Religious Education, 32*(2), 93–103. doi:10.1080/01416200903537399.

United Nations Environment Programme (UNEP). (2015). UNEP environmental, social and economic sustainability framework, January 2015. http://www.unep.org/about/eses/Portals/50272/Documents/UNEP_Environmental_Social_and_Economic_Sustainability_Framework.pdf

Wall, J. (2010). Ethics in the Light of Childhood, Washington D.C.: Georgetown University Press.

Philosophizing with Transdisciplinarity, Relational Knowledge and Ethics in Education for Sustainable Development

Marie Grice

Introduction

This chapter sets out to pursue and explore the concepts of transdisciplinarity and collaboration, particularly in education for sustainable development (ESD). This is done by providing a conceptual analysis that will allow relational and ethical aspects to surface where individual ones generally rule. As its starting point, this chapter takes a practical educational context and an empirical study of students involved in a book project regarding sustainable development. The aim of the study was to quantitatively explore what concepts of knowledge the students might display and how these impacted on how the students oriented themselves in such an educational context (Grice 2014). The object of study was a transdisciplinary educational project concerning various sustainability-related issues regarding food—for example, production, distribution, consumption, starvation and waste. About 300 Swedish upper-secondary students from a number of schools participated. Provided with contact details to 40 available extramural experts in various fields, the students explored their topics and produced minichapters for the

M. Grice (✉)
Department of Technology, Uddevalla Upper Secondary School,
Uddevalla, Sweden

O. Franck, C. Osbeck (eds.), *Ethical Literacies and Education for Sustainable Development*, DOI 10.1007/978-3-319-49010-6_2

book. Two of the concepts of knowledge emerging from the factor analysis in the study were identified as transdisciplinary knowledge and collaborative knowledge. Both proved significant in how the students evaluated the educational project for sustainable development that they were involved in. As independent variables in a multiple regression analysis, the epistemic beliefs had a considerable impact compared to background variables such as gender, age and study programme. For this reason it seems relevant to further analyse and explicate what notions such epistemic beliefs might comprise. In a broad sense the results of the empirical study are the rationale for pursuing notions of transdisciplinary education, relational knowledge and ethics in this chapter.

Method

Here the topic of transdisciplinarity, relational knowledge and ethics in ESD are approached from an eclectic, theoretical angle. The research process has abductive traits in that various theoretical frameworks and lenses inspire the discovery of new patterns and create understanding. By tracing concepts such as transdisciplinarity and relational knowledge, through relevant research literature, using various search strings such as transdisciplinarity, transdisciplinarity and education, and transdisciplinarity and ESD, numerous research areas were found. In choosing the relevant research, selection was carried out in a snowball fashion, beginning with the literature used in a previous study (Grice 2014). Reference lists and search strings provided further literature, adding converging and diverging aspects regarding transdisciplinarity, relational knowledge and ethics. When the variety of new perspectives seemed to be saturated and the same aspects kept repeating themselves, it was decided that the literary search was sufficient.

Exploration of the concepts of transdisciplinarity and transdisciplinary education was further guided by the application of Dohn's method of "philosophizing with" (2011). Dohn recognizes four ways that epistemology can engage with knowledge sciences. The first three—conceptual analysis, conceptual clarification and implication, and scientific result interpretation—are recognized as conventional applications of philosophy. Dohn introduces a fourth: epistemology as a dialogue partner with a voice of its own. This approach goes beyond conventional applications of philosophy (ibid.). It is used here as an analytical lens in the epistemic exploration from concept to educational practice.

To provide an open and critical approach to transdisciplinary inquiry in ESD, the theory of knowledge production needs to be investigated. Therefore an account of transdisciplinary knowledge needs to pay attention to ontological, epistemological and ethical aspects (Russell 2010). Another lens applied in the abductive approach is Nancy Tuana's conceptualization of moral literacy, involving three basic components: ethics sensitivity, ethical reasoning skills and moral imagination (2007). This framework translates the theoretical elaborations to the practical applications in the classroom and provides a strong pedagogical strand, suggesting that moral literacy is a practical concept which is both the goal and the means to develop ethical competence. Indeed, the basic components of moral literacy might be viewed as core competences in general education and specifically in sustainability education. Together with imagination and sensitivity, reasoning skills will add important aspects to the learning process, to knowledge creation for sustainable development and also to education in general.

Epistemic Beliefs and Education for Sustainable Development

The concepts of knowledge, epistemic beliefs, fall into the category that the educational psychologist calls personal epistemology, a construct that refers to how students understand knowledge and how they come to know (Hofer 2004). In a number of studies, personal epistemology has been related to comprehension, learning and education. Epistemic beliefs and education seem to affect each other interchangeably. Despite issues regarding terminology and methodology (ibid.), there are indications that epistemic development in this sense might be signified by critical thinking, problem solving and the relationship between theory and evidence (ibid.). Such aspects can also be found in the conceptualizations of competences associated with transdisciplinary research (Hirsch Hadorn et al. 2006) and ESD (Madsen 2013).

In an educational response to the global issues of sustainable development, fundamental questions need to be asked, such as what knowledge and knowing might be. As the two emerging concepts of knowledge or epistemic beliefs, in the previously mentioned study, transdisciplinary knowledge and collaborative knowledge, carry the notion of relationships, such aspects are worth additional investigation. Thus the overall aim of this chapter is to further explore, or *philosophize with*, aspects of transdisciplinarity and collaboration in ESD. In educational terms and in society, there may not be a clear distinction between various conceptions of knowledge. An interest in the chapter is whether epistemology can help to make us see various concepts of knowledge in educational practice and curriculum in order to deal with complexity and uncertainty in research and ESD. However, it is assumed here that understanding knowledge in terms of analytical philosophy is not the purpose of philosophizing with, but by raising awareness of issues in epistemology, relevant concepts of knowledge can be detected. By focusing on relational aspects in the knowledge-creation process in ESD, epistemological issues might surface regarding what knowledge is necessary for the individual and the collective to develop sustainability knowledge and ethics.

Relational Conceptualizations of Education for Sustainability

At the core of ESD there are at least three aspects of sustainability, which relate to environmental, societal and economic dimensions. Translated into the pillar metaphor, these can be viewed in a generic way as being centrifugal and global. In various sustainability frameworks a fourth pillar has been suggested (Burford et al. 2013)—for example, a cultural-aesthetic, an institutional-political or a spiritual-religion dimension. The United Nations Educational, Scientific and Cultural Organization (UNESCO) has supported the cultural aspect of sustainable development and is a strong advocate of such a missing pillar. However disparate these conceptualizations may seem, they share a concern for contextualization and ethical values (ibid.). The fourth pillar seems to be connected to the local and the particular, and it might correspond to the need to couple the global and local perspectives, recognizing the "glocal" level of knowledge production (Bäckstrand 2003). These perspectives are understood as interconnected and being in dialogue, which provides a basis for relational knowledge and ethics. As there is no given end or solution to the challenges of sustainable development, part of the solution will be to incorporate the element of uncertainty among the concepts of knowledge.

In a society frequently referred to as complex, dynamic and changeable, it seems relevant to ask fundamental epistemic questions and to philosophize *with* a concept such as transdisciplinarity. While effort has been made regarding pedagogical techniques employed to promote participation and higher-order thinking (Mochizuku and Fadeeva 2010), less energy has been expended regarding what to assess in inter- or transdisciplinary education. Before such an attempt can be made, a clear understanding of transdisciplinarity and core competences is required (Boix Mansilla et al. 2009).

Relational and Epistemological Aspects of Knowledge and Knowing in Education for Sustainable Development

The global call for sustainable development is society's response to so-called "wicked problems" (Russell 2010), which are characterized by conflicting scientific theories and evidence or "megaproblems" (Warwick 2012), which relate to the magnitude and geographical scope of global and intergenerational threats to the ecosystem. In coming to terms with wicked problems and global challenges, transdisciplinary approaches are suggested. From the complexity of the problems follows that one science or discipline will not be able to solve them fully (Osorio et al. 2009). As we know from history, one solution to a problem can give rise to another issue if the problem is not dealt with in a holistic and comprehensive way. For example, technical solutions to the problems need to be sensitive to the planet and people, even if they only set out to deal with very local and contextualized issues. This adds complexity to both the problem definition and the problem solution. Systems thinking and holism are theories that underpin the transdisciplinary approach to the solution of the global challenges of sustainable development. Systems thinking in science corresponds to the notion of interrelatedness that is apparent in many conceptualizations of sustainable development (Summers et al. 2005; Osorio et al. 2009). The interconnectivity of decisions and actions relates to systems thinking and will also give rise to questions about how the human system is to be understood and related to the natural system. For the individual mind it seems more or less futile to grasp the various systems that they might belong to, affect and be affected by. The human system equated with the technological system can be viewed as part of, inside or outside the natural system. Depending on the prioritized interpretation, various ethical issues concerning the relational processes will surface. These issues might concern who, what and how responsibility should be distributed between, within or beyond systems.

Holism in transdisciplinarity is an approach that looks for coherence rather than unity of the system, the object of study (Thompson Klein 2004). It is a concept that includes and allows for multidimensionality. New knowledge models have been suggested to transcend the boundaries of disciplinarity, which in our modern complexity conception of the world has been deemed wanting. Key to these models is the integration of knowledge and methods from various disciplines. By inviting knowledge and resources from multiple disciplines and various stakeholders from society, and with a democratic agenda, laypeople as well as experts are seen as important contributors to solving the problems (Bäckstrand 2003; Hirsch Hadorna et al. 2006). Transdisciplinarity and other integrative approaches have been proposed as an ideal, without recognizing the backwash of conflicts and disharmonies in practice (Klenk and Meehan 2015). In the wake of the vision of integrated knowledge comes the power and politics of knowledge, which might in fact limit the possibilities of transdisciplinary problem solving and knowledge creation (ibid.). Whether it be research or education, the transdisciplinary approach might be seen as a futile one if the aim is to comprise a complete understanding of the problem and a consideration of all possible methods that might be used to solve the sustainability challenge, involving all possible stakeholders. Indeed, we most likely would not have a problem if such an attempt was doable. These considerations of consequences beyond the actual problem can be seen to move horizontally and centrifugally from the problem to the periphery. The gyration itself can be related to the transdisciplinary method and systems thinking in that the problem is of such proportions that it sweeps across a variety of disciplines. No single discipline will be able to provide its solution. With its holistic perspective, transdisciplinarity aims to consider all possible aspects of a problem as well as approaches to its solution. The possibility of such a methodology has been questioned (Kopnina and Meijers 2014) and consequently poses both epistemological and ethical concerns for educational practice.

With an awareness of the complications that follow the holistic approach, it might be productive to consider creative holism, which nurtures a critical systems approach to complex problem situations (Jackson 2006). Creative holism includes learning about and harnessing various systems methodologies and practices (ibid.). It shares with critical systems thinking a commitment to critical awareness, pluralism and improvement. An intervention will follow four phases: creativity, choice, implementation and reflection. The methodology of systems thinking which is suggested in the literature (Angelstam et al. 2013), necessary for solving various socioscientific issues, requires new competences and relationships among the workforce and general public, including students. It requires that people see themselves as parts of different systems and it requires ethical sensitivity in all four phases. If sustainability is seen as a knowledge system, the human being is part of and beyond it but not in a hegemonic manner. The sense of place for the human being as part of, or beyond, rests on how the systems are defined and entangled with other systems, such as the natural system, without any hegemonic order between them. Developing knowledge about systems and their interrelatedness and considering one's own sense of belonging or lack of belonging to defined systems will be both an epistemic and an ethical objective of ESD.

Relational Ethics in the Transdisciplinary Approach to Sustainable Development

In ESD, real-world education addresses the complexity of problem definition and solution. The four-pillar framework—economic, social, ecological and cultural—corresponds to systems thinking. Understanding that systems are made up of parts or individuals that depend on each other could make students see themselves as belonging to various systems. All parts, individuals and non-human entities express the interrelatedness of the various components of the system. Such systems thinking in education can provide common ground for ethical values and ESD competence. Awareness of the effects of the systems could be a necessary ethical driver in critical reflection on our responsibility and moral obligations within and between systems.

Key learning processes in ESD comprise collaboration and dialogue, engagement with the whole system, and active and participatory learning (Cebrián and Junyent 2015), all of which place relationship as more important than the individual. Therefore individual-centred education does not correspond to the challenge of sustainable development and current demands of teamwork and collaboration (Gergen 2011). The sustainability competences that need to be developed are problem solving, critical thinking, action competence and systems thinking (Varga et al. 2007; Cebrián and Junyent 2015). The concept of competence seems to be in use by various international bodies, such as the Europen Union and the UN, which have developed various frameworks (Cebrián and Junyent 2015). There are both converging and diverging conceptualizations of the concept of competence associated with ESD (Mochizuku and Fadeeva 2010). Among educators the concept of competence is viewed with some suspicion because of the obvious risk of oversimplifying the epistemic goals of education. One of the arguments put forward by Claxton (2009) is that complex performances are divided into component skills, which could be developed or trained in education. Consequently it is assumed that such cognitive subroutines (p. 183) are reassembled in the individual learner. Another limitation of the competence conceptualization of learning outcomes is the assumption that the training of various cognitive subroutines can rather rapidly create a mental capacity capable of constructing and evaluating explicit, reasoned argument.

Critical thinking and reasoning have a strong link to moral literacy. Nancy Tuana argues that moral literacy is a skill that should be fostered in school as our complex society exposes children and students to various ethical issues, large and small, such as cheating, bullying and refraining from taking action (e.g. failing to stop a friend who has been drinking from driving their car; Tuana 2007). In Tuana's conceptualization of moral literacy, it is a skill that needs to be fostered and developed over time and in school by teachers who are well acquainted with moral subject matter. Compared with the importance devoted to literacy in the contexts of mathematics, languages, science and reading, attention to moral literacy is scarce if not absent. Tuana uses the literacy concept for three reasons. First, moral literacy is a complex concept that requires various advanced competences and skills. Second, it is seen as a continuum that pupils and students could develop more or less of. Third, it is a skill that education could foster.

In Tuana's conceptualization, moral literacy is made up of three components: ethical sensitivity, ethical reasoning skills and ethical imagination. Ethical sensitivity is at play both in the identification of the issue—whether it is to be considered as ethical or not—and in the judgement of the intensity of the ethical dimension of the issue. The same component is active in the choice of action to be taken in response to the issue. The skills needed to develop ethical sensitivity can be included in the curriculum across the whole school system, at all levels. In ESD there are various issues that provide contexts for considering ethical responsibility for the consequences of actions. Moral literacy can be seen as one element of sustainability competence, the knowledge underpinning good decisions. More important is what practical manifestations are made in the name of moral literacy. How do we get from knowing to doing what we know is right? Here, action competence, the willingness to be a competent participant (Jensen and Schnack 1997), plays a role in relation to phronesis, an intellectual virtue resting on deliberation based on values and concerned with practical judgement (Kinsella and Pitman 2012). Yet another competence concept, ethical competence, has been suggested: action readiness (Grice and Franck, 2017). In addition to moral literacy and action competence, the concept of action readiness includes the aspect of not refraining from taking action. Owing to the limited space here, the line of research regarding why individuals avoid decisions by postponing them, failing to act and accepting "business as usual" (Anderson 2003)—the psychology of doing nothing—cannot be developed here, but it is suggested for further research. Moral motivation (Kaplan and Tivnan 2014) offers further conceptualizations regarding the self-organization of cognitive and emotional dynamics, which might be worth pursuing to learn how moral judgement can be turned into action.

In one conceptualization of the general aims of ESD it has been suggested that the students should:

- be given a chance to orientate themselves among existing viewpoints and opinions;
- achieve knowledge and ethical awareness in order to be able to critically evaluate alternatives;
- develop an ability to take action regarding sustainable development issues;
- be invited to participate in activities that reveal the meaningfulness of engaging in sustainability issues (Sandell et al. 2005).

These aims tally with the suggested orchestration of ESD in this chapter. Transdisciplinary knowledge, relational ethics and action readiness are the epistemic beliefs and competences that seem to be what is required of students to cope with conflicts and complexities in this context. By engaging in real-world issues and interacting with the community regarding sustainability issues, students will not be allocated a marginalized status in the community of citizens. In policy documents and curricula, students are sometimes placed in the margins of citizenship practice and depicted as lacking in knowledge, competences and values (Olson et al. 2014). By contrast, in the transdisciplinary approach, diverse forms of knowledge are called for and students as stakeholders will be recognized as co-producers of knowledge.

Independent Self and Relational Being

Transdisciplinarity is clearly linked to a holistic worldview, which recognizes the connectedness between individuals. Transdisciplinarity and relational being make up a critique against the individualistic society and current single-disciplinary method of problem solving. Values are founded in relatedness, and this relatedness provides a context for morality (Miller 1986). Freedom and responsibility can be seen as fundamental components of such independency and relationships.

It is important to highlight the concept of freedom both for the identity of the individual and for an understanding of the individual in and of the collective. What sort of freedom can the collective appreciate? Freedom to do and freedom to be? What political and social freedom can be given to the individual considering their relation to other people and animals, plants and the planet? How can we set limits to individual freedom in the name of sustainable development? Rather than focusing on the limitations of freedom and problems in relation to sustainability, we need to discuss what freedom could be in a society that rests on sustainable ethics.

Freedom is a fundamental value put forth in, for example, the Swedish curriculum for compulsory schooling. Under the heading "Fundamental values", expressions of the inviolability of human life; individual freedom and integrity; equality for all, men and women alike; and a solidarity with the weak and vulnerable can be found (Skolverket 2011, p. 9). Furthermore, pupils are encouraged to acknowledge their uniqueness and to "participate in the life of society by giving of their best in responsible freedom" (ibid.). As previously pointed out by Grice and Franck (2014), it might not be pedagogically responsible to support individualistic features in the action for sustainability competence. Rather, one might want to think about this in terms of collective, interdependent or even related action, which needs to display elements of consideration and sensitivity of the other, not only the individual other but also the global community other and the physical world other, that are around us, including all animals and plants (Amerigo et al. 2012). Instead of using the self, whether it be I or you, as a unit of analysis, a relational approach would suggest that we use the relationship itself as the unit of analysis (Salipante and Koury King 2010). However, the two units of analysis seem to belong to two categories: one an object, a socially constructed self; the other a process or rather the self set in motion, relationally.

The Educative Moment: *Le Moment* in an Educational Context

A great many have argued for the necessity of transformative education to address global challenges (Pavlova 2013). "It requires recognizing the interconnectedness among universe, planet, natural environment, human community, and personal world through critical reflection, holistic approaches and relationships with others" (p. 735). This transformation is rooted in an educational philosophy that identifies ethical development as a key objective of education. This ethics concern valuing the other person, moral responsibility and the recognition of a non-instrumental relationship with nature. Pavlova suggests that a transformative pedagogy should:

- help students to recognize a situation as being ethically (morally) problematic;
- enable students to have a voice and express their feelings and thoughts;
- find a solution that serves the best interests of all parties involved (p. 741).

The transformative pedagogy resonates well with Tuana's moral literacy. Ethics sensitivity is one of the three basic components of moral literacy. It is the ability to determine whether a situation involves ethical concerns. The other two components—ethical reasoning skills and moral imagination—correspond to expressing feelings and thought, and finding an ethical solution, respectively.

One of the characteristics of transdisciplinarity is its centrifugal, problem-driven movement. It is what reveals itself in this gyration that can be captured by the teacher as an educative moment in addition to other predetermined parts of ESD. The educative moment, here referred to as *le moment*, might reveal itself as an opportunity that cannot be predicted or determined but needs to be sensed and seized. A creative quality and moral disposition of the teacher is to be sensitive to *le moment*. I have borrowed the French term *le moment* from the doctrines of the naturalistic nineteenth-century literary era, which was underpinned by determinism. Émile Zola, the writer of the naturalistic manifesto, claimed that the process of the writer was similar to that of the scientist. He based his views on Hippolyte Taine, a thinker and literary critic, who could determine a writer according to the three elements of *la race, le milieu* and *le moment* (*Encyclopaedia Britannica* 2016).

Taine sees *le moment* as historic time, and claims that the creation of the writer is determined by inner and outer factors at a specific point in time (Nias 2013). *Le moment*, as used in this chapter, in an educative moment is rather inspired by the practical application of the faculty of *le moment* in a naturalistic drama such as *Miss Julie* by Swedish author August Strindberg. The biological heritage of the heroine (race) in combination with her upbringing without a mother (milieu) can indeed be seen as determining factors that led to the inescapable, tragic end where she commits suicide. However, in the play, *le moment* can be seen as a creative opportunity or chance—the seductive moment of the midsummer night's dance and music—which sets the whole process into play. Without *le moment*, *la race* and *le milieu* would not come into effect. It is this element of unpredictability despite the author's intention to explain his characters through heredity and environment that makes *le moment* a workable concept in the description of educative moments and spaces in ESD. It is its creative dimension that corresponds to what is required by the teacher in grasping the opportunity to orchestrate learning.

Given that knowledge is understood as co-created and education a process of participation in a relational process, new pedagogical practices might emerge (Gergen 2011). Education as a relational process for participating in sustainability practice requires a teacher who recognizes no epistemological or pedagogical barriers in school or out of school in interacting with students, teachers, office staff or other stakeholders in society. The demand on knowledge in the transdisciplinary approach is that it is problem focused, relevant and communicable among various collaborating stakeholders (Russell et al. 2008). To meet this demand the teacher needs to actively take their students out into society and work on various sustainability issues. It is important that the students get the experience of working for sustainable development together with their teacher. Learning together with the teacher, other students and stakeholders in sustainability activities might lead to learning how to participate. As Gergen suggests' "the primary aim of education is to" enhance "the potentials for participating in relational processes—from the local to the global" (2011, p. 243). It is suggested here that to enhance "the potentials", ethical development is necessary.

Transdisciplinary learning—in real life and involving several stakeholders—particularly allows for sharing skills and experience, and creating new knowledge (Park and Son 2010). In this context it seems relevant to address concepts such as relational responsibility and relational ethics.

It is suggested here that to understand and facilitate the learning process in ESD, the teacher has to understand the role of ethics in the relational co-creation of knowledge in the transdisciplinary mode. One way to do so could be to develop the element of care in education. Shared care makes the educational moment one of mutual responsibility, and a form of boundary crossing takes place between the teacher and the student as the traditional roles of teacher and learner are to some extent switched. Learning together with one's teacher is a relational process that requires some development of epistemological strands of thought or epistemic beliefs. The process of collectively creating knowledge provides an opportunity for *le moment* to take place. It takes sensitivity and readiness, and the willingness of all the participants to act in favour of, for example, sustainable development. The individual becomes aware of and realizes their relational spaces, individually and collectively. The learning outcome of such a knowledge-creating orchestration of education might be referred to as relational knowledge.

On both a motivational and an epistemic level, it is relevant to look at boundary crossing in transdisciplinary education and research. What is the boundary crossing itself but a moment of epistemic development for the individual or the group/collective. Perhaps boundary crossing is not exclusive to the transdisciplinary approach and perhaps it is even irrelevant to think of any knowledge creation as not boundary crossing. In all knowledge creation it is necessary to be sensitive to these moments of boundary crossing or transgression because they are educative moments. For the learner it is a question of both seeing and becoming aware of the boundary—being put in the boundary, in the sphere of the boundary and being aware of the educative moment. *Le moment*, although it seems to have a time reference, could also be conceptualized as a space. It is a time-space perspective that the teacher needs to recognize in order to be sensitized to the knowledge creation. Indeed, the origin of the word "moment" is the Latin word *momentum*, the root of which is *movere*—to move (*Oxford English Dictionary*). What is creation if not a moment in time when knowledge comes to be and a learner comes to know. Knowledge creation can be conceived of as a quick moment of coming to know, such as "a moment ago I did not know", versus "a moment later I knew".

On a motivational level, the personal epistemology that one needs to cross or straddle or simply include this middle, this in-between, might be related to one's epistemic beliefs. It might be possible to think of epistemic beliefs as the reasons why some students/learners will or will not see reasons to epistemically, logically, ethically or practically cross that boundary.

It is possible that the individual knower will experience this instantaneous change in knowledge. At the same time, in transdisciplinarity there is a focus on a group solving a problem and thereby coming to know. The focus in transdisciplinary knowledge creation is on different stakeholders in collaboration, solving an issue or problem. These are two different processes in one. However, there is no guarantee that the whole group will learn the same thing. The various stakeholders' learning processes will be different. Transdisciplinarity could simply be seen as a tendency and willingness to connect with other knowers, knowledge and possible solutions. It is underpinned by a belief in constructed knowledge, and that it is possible for the collective society to improve. It gives hope and it is a choice—to be ready to change and be transformed, knowing through being and doing.

Conclusions

This chapter has raised epistemological, relational and ethical issues with regard to ESD and transdisciplinarity. For educational practice transdisciplinarity in education regarding authentic, real-world problems in a joint venture with stakeholders out of school, should prepare students to deal with epistemological variety and uncertainty. However, relational or interpersonal activities in education must not be limited to the organization of education in which learning is taken for granted or somewhat naïvely presumed to take place with the right orchestration of the task. It is also important to open up this sort of education for values and moral education. I illustrate my claims by arguing for the important contributions that relational aspects of knowledge and ethics can make to ESD, and I contend that moral literacy can be honed by and emerge from transdisciplinary efforts in education, including collaboration in real-world situations. *Philosophizing with* is a dialogic application to let epistemology give voice in issues of ESD, transdisciplinarity, competences, action readiness and moral literacy. The concept of *le moment* was introduced to highlight the educative moments that might emerge when dealing with uncertain knowledge in transdisciplinary modes. A variety of knowledge is valued in transdisciplinarity. In educational practice it takes the epistemological awareness and ethical sensitivity of the teacher to identify and grasp the educative moment where the mode of learning by the student is illustrated by the concept of *le moment*.

References

Amerigo, M., Aragonés, J. I., & García, J. A. (2012). Exploring the dimensions of environmental concern. An integrative proposal. *Psyecology, 3*(3), 353–365.

Anderson, C. J. (2003). The psychology of doing nothing: Forms of decision avoidance result from reason and emotion. *Psychological Bulletin, 129*(1), 139–167.

Angelstam, P., Andersson, K., Annerstedt, M., Axelsson, R., Elbakidze, M., Garrido, P., et al. (2013). Solving problems in social–ecological systems: Definition, practice and barriers of transdisciplinary research. *Ambio, 42*(2), 254–265 DOI 10.1007/s13280-012-0372-4.

Bäckstrand, K. (2003). Civic science for sustainability: Reframing the role of experts, policy-makers and citizens in environmental governance. *Global Environmental Politics, 3*(4), 24–41.

Boix Mansilla, V., Dawes Duraisingh, E., Wolfe, C. R., & Haynes, C. (2009). Targeted assessment rubric: An empirically grounded rubric for interdisciplinary writing. *The Journal of Higher Education, 80*(3), 334–353.

Burford, G., Hoover, E., Velasco, I., Janoušková, S., Jimenez, A., Piggot, G., & Harder, M. K. (2013). Bringing the "missing pillar" into sustainable development goals: Towards intersubjective values-based indicators. *Sustainability, 5*, 3035–3059.

Cebrián, G., & Junyent, M. (2015). Competencies in education for sustainable development: Exploring the student teachers' views. *Sustainability, 7*, 2768–2786.

Claxton, G. (2009). Cultivating positive learning dispositions. In H. Daniels, H. Lauder, & J. Porter (Eds.), *Educational theories, cultures and learning: A critical perspective* (pp. 177–187). New York/Routledge: Abingdon/Oxon.

Dohn, N. B. (2011). Roles of epistemology in investigating knowledge: 'Philosophizing with'. *Metaphilosophy, 42*(4), 431–450.

Gergen, K. J. (2011). *Relational being: Beyond self and community.* New York: Oxford University Press.

Grice, M. (2014). Epistemic beliefs and knowledge creation among upper-secondary students in transdisciplinary education for sustainable development. *Nordidactica, 1*, 146–169.

Grice, M., & Franck, O. (2014). A phronesian strategy to the education for sustainable development in Swedish school curricula. *Journal of Education for Sustainable Development, 8*(1), 29–42.

Grice, M., & Franck, O. (2017) Conceptions of ethical competence in relation to action readiness in education for sustainable development. *Reflective Practice, 18*(2), 256–267, DOI: 10.1080/14623943.2016.1269001

Hirsch Hadorn, G., Bradley, D., Pohl, C., Ristd, S., & Wiesmann, U. (2006). Implications of transdisciplinarity for sustainability research. *Ecological Economics, 60*(1), 119–128. doi:10.1016/ j.ecolecon.2005. 12.002.

Hofer, B. K. (2004). Introduction: Paradigmatic approaches to personal epistemology. *Educational Psychologist, 39*(1), 1–3.

Jackson, M. C. (2006). Creative holism: A critical systems approach to complex problem situations. *Systems Research and Behavioral Science, 23*, 647–657.

Jensen, B. B., & Schnack, K. (1997). The action competence approach in environmental education. *Environmental Education Research, 3*(2), 163–178.

Kaplan, U., & Tivnan, T. (2014). Moral motivation based on multiple developmental structures: An exploration of cognitive and emotional dynamics. *The Journal of Genetic Psychology, 175*(3), 181–201.

Kinsella, E. A., & Pitman, A. (2012). Phronesis as professional knowledge: Practical wisdom in the professions. In E. A. Kinsella & A. Pitman (Eds.), *Phronesis as professional knowledge: Practical wisdom in the professions /* [Elektronic resource] (pp. 1–11). Rotterdam/Boston: Sense Publishers.

Klenk, N., & Meehan, K. (2015). Climate change and transdisciplinary science: Problematizing the integration imperative. *Environmental Science & Policy, 54*, 160–167.

Kopnina, H., & Meijers, F. (2014). Education for sustainable development (ESD). *International Journal of Sustainability in Higher Education, 15*(2), 188–207. doi:10.1108/IJSHE-07-2012-0059

Madsen, K. D. (2013). Unfolding education for sustainable development as didactic thinking and practice. *Sustainability, 5*(9), 3771–3782.

Miller, J. P. (1986). Atomism, pragmatism, holism. *Journal of Curriculum and Supervision, 1*(3), 175–196.

Mochizuku, Y., & Fadeeva, Z. (2010). Competences for sustainable development and sustainability. *International Journal of Sustainability in Higher Education, 11*(4), 391–403.

Nias, H. S. (2013). Hippolyte Taine (1828–1893). In M. A. Habib (Ed.), *Cambridge history of literary criticism* (pp. 393–405). Cambridge: Cambridge University Press.

Olson, M., Fejes, A., Dahlstedt, M., & Nicoll, K. (2014). Citizenship discourses: Production and curriculum. *British Journal of Sociology of Education, 36*(7), 1036–1053.

Osorio, L. A., Lobato, M. O., & Del Castillio, X. Á. (2009). An epistemology for sustainability science: A proposal for the study of the health/disease phenomenon. *International Journal of Sustainable Development & World Ecology, 16*(1), 48–60.

Park, J.-Y., & Son, J.-B. (2010). Transitioning toward transdisciplinary learning in a multidisciplinary environment. *International Journal of Pedagogies and Learning, 6*(1), 82–93.

Pavlova, M. (2013). Teaching and learning for sustainable development: ESD research in technology education. *International Journal of Technology and Design Education, 23*(3), 733–748.

Race, milieu, and moment. (2017). In Encyclopædia Britannica. Retrieved June 12, 2016, from http://academic.eb.com.ezproxy.ub.gu.se/EBchecked/topic/488048/race-milieu-and-moment

Russell, J. Y. (2010). A philosophical framework for an open and critical transdisciplinary inquiry. In J. Harris, V. A. Brown, & J. Russell (Eds.), *Tackling wicked problems: Through the transdisciplinary imagination* (pp. 31–60). London: Earthscan.

Russell, A. W., Wickson, F., & Carew, A. L. (2008). Transdisciplinarity: Context, contradictions and capacity. *Futures, 40*(5), 460–472.

Salipante, P., & Koury King, N. (2010). Chapter 5. The heart of relational organizing: Passion, autonomy and responsibility. In C. Steyaert & B. Van Looy (Eds.), *Relational practices, participative organizing* (pp. 77–102). Bingley: Emerald Group Publishing Limited.

Sandell, K., Öhman, J., & Östman, L. (2005). *Education for sustainable development – Nature, school and democracy.* Lund: Studentlitteratur.

Skolverket [Swedish National Agency for Education]. (2011). *Curriculum for the compulsory school, preschool class and the leisure-time centre.* Stockholm: Skolverket.

Summers, M., Childs, A., & Graham, C. (2005). Education for sustainable development in initial teacher training: Issues for interdisciplinary collaboration. *Environmental Education Research, 11*(5), 623–647.

Thompson Klein, J. (2004). Prospects for transdisciplinarity. *Futures, 36*(4), 515–526.

Tuana, N. (2007). Conceptualizing moral literacy. *Journal of Educational Administration, 45*(4), 364–378.

Varga, A., Kószó, M. F., Mayer, M., & Sleurs, W. (2007). Developing teacher competences for education for sustainable development through reflection: The environment and school initiatives approach. *Journal of Education for Teaching, 33*(2), 241–256.

Warwick, P. (2012). Climate change and sustainable citizenship education. In J. Arthur & H. Cremin (Eds.), *Debates in citizenship education* (pp. 132–145). London/New York: Routledge/Taylor & Francis Group.

Pupils' Views on Moral Competence in School

Annika Lilja

Introduction

Teachers often testify that during lessons, pupils reason wisely about how to act ethically and how to differentiate between good and bad, right and wrong. By contrast, when it comes to handling the relationships between each other, wise reasoning does not seem to be applied by all of them (Beem et al. 2004; Lilja in press). The actions of pupils bullying or abusing each other often depend on ingrained patterns that are part of the existing culture of the school, and because of that they happen in an unreflecting way (Skolverket 2011b).

An overall objective of school in Sweden is to foster pupils' abilities to develop knowledge in different subjects, but also foster values such as equality, solidarity, human rights and democracy. This can be found in the general guidelines in the curriculum for compulsory school. Another aim is to give pupils opportunities to describe and analyse ethics as a field of knowledge. This is described in the syllabus for religious education (RE; Skolverket 2011a). These two objectives can be said to support the development of ethical literacy.

A. Lilja (✉)
University of Gothenburg, Gothenburg, Sweden

O. Franck, C. Osbeck (eds.), *Ethical Literacies and Education for Sustainable Development*, DOI 10.1007/978-3-319-49010-6_3

It is hard to reach a state where all pupils feel comfortable and safe in school. Even though it is stated in the curriculum that all pupils are to respect other people's intrinsic values (p. 12), most teachers struggle in different ways to create and maintain sustainable relationships to make all pupils feel included, and some pupils are bullied, teased or abused every day in most schools.

The aim of this study is to highlight pupils' perspectives on issues regarding ethical literacy and sustainable relationships. In interviews, 11 pupils, five aged 11 and six aged 15, were asked why some pupils tease or abuse their classmates. The participants were asked what they think their school can do to prevent bad behaviour between pupils. This work focuses on sustainable relationships in a Swedish context, but according to previous research (e.g. as described below) there are reasons to believe that the pupils' views are relevant from an international perspective.

Previous Research

There is a lot of research about bullying, both from the bullies' and from the victims' perspectives (Varjas et al. 2008). Moreover, work on children's and adolescents' perspectives on bullying is increasing (Forsberg 2016). Most of these studies are about bullying in a school context (Bibou-Nakou et al. 2012). When pupils are asked who becomes a victim, several investigations present the same answer (e.g. Forsberg 2016; Varjas et al. 2010): the victims are described as odd or different, and insecure (Frisén et al. 2008; Forsberg 2016); or they provoke the bullies (Frisén et al. 2008). An explanation for why some pupils bully and abuse other pupils is that the bullies strive for power and status (Thornberg 2010; Forsberg 2016). Forsberg (2016) established in her thesis that social ordering and belonging seem to be very important in pupils' social lives, and when negotiating these relationships, bullying appears to play a crucial role. She found three subcategories of social ordering: social hierarchical ordering, peer ordering and new member ordering.

In schools where pupils are engaged and are allowed to participate in decisions that concern them, grades are higher, social sustainability is stronger and bullying is less widespread than in schools where pupils are not allowed to participate in such decisions (Ahlström 2009). Another way to prevent pupils from teasing and abusing each other is for the school to establish structures that offer emotional support to vulnerable pupils and also give befrienders the opportunity to find practical ways to demonstrate empathy for those who are exposed (Jennifer and Cowie 2012). Another strategy, according to Jennifer and Cowie (2012), is to improve daily life in the school by enabling children to explore the emotions aroused in situations where someone is the subject of some form of indignity. An exercise like this will give the pupils the opportunity to reflect on how feelings such as shame and guilt arise.

In a report from Skolverkert (2011b) the researchers state that a lot of non-bullying programmes used by schools in Sweden focus on how to make pupils understand the victim in a bullying situation. The schemes are not simply about understanding who is "in" and accepted and who is "out" and a victim. The researchers further report that the exercises that teachers and pupils carry out in these programmes are often general and not specifically tailored to different situations relevant to the specific group of pupils. The fact that the exercises are general makes it hardert for the pupils to be engaged and hard to incorporate and adopt a different behaviour. Another problem highlighted by the report is that the exercises are only about situations in the classroom. During the breaks, in the corridors and the school restaurant, other rules seem to apply. Jennifer and Cowies' study (2012) "indicates the importance of working with peer group relationships as a fundamental way of addressing the issue of bullying" (p. 238).

Creating and developing relationships that are sustainable over time, both with teachers and with other pupils, is necessary to prevent pupils from abusing each other (Ahlström 2009). Elvstrand (2009) also found in her thesis about 11-year-old pupils and their participation in school that building relationships is important. Both teachers and pupils are engaged in this but in different ways. The teachers in Elvstrand's study had an explicit idea of how to create solidarity in the group and how to interact with each individual pupil. The pupils sometimes had problems with building relationships with other pupils so they needed appropriate support. Without such support, relationships can end in conflict. Trust is also an important part in a sustainable relationship (Lilja 2013). A trusting relationship between a teacher and their pupils implies that it has a stable foundation from which both teacher and pupils are able to manage the different challenges that the relationship involves (ibid.).

The objective of working with ethical literacy and supporting sustainable relationships, both between pupils and between a teacher and their pupils, is something that all teachers are responsible for. In Sweden, ethics is part of the syllabus of RE, so those teaching the subject have the opportunity to work with existential phenomena such as alienation and violation during

RE classes. Pupils recognize and engage in lessons about these phenomena (Osbeck in press), and, as a consequence, such work can serve two purposes: to enhance the development of ethical skills and to prevent violations between the pupils. By developing the students' knowledge about alienation and violation, their linguistic repertoire regarding these phenomena can increase both when it comes to interpreting life and when it comes to informative linguistic improvement. Stories that interpret life might create an experience of and an insight into these questions about alienation and violation, and from different perspectives. Work with ethical literacy by using stories in this way, pupils get opportunities both to recognize their own perspective and to realize that there are alternative ways to experience a situation (ibid.).

The research reported above aims to give a background to why pupils abuse their classmates, and what has been determined as successful courses of action for schools when it comes to supporting pupils' ethical literacy and sustainable relationships between pupils.

Theory

To interpret and understand the pupils' answers regarding why some choose to abuse and tease classmates while some stop themselves from doing so, and to understand what the pupils suggest that the teachers at their schools can do to prevent bullying, the theories of Martha C. Nussbaum and Knud E. Lögstrup were chosen. Nussbaum's theories were applied to discuss how the pupils want the school to work with ethical literacy—for example, using values such as equality, solidarity, human rights and democracy. Lögstrup was chosen to elaborate on the pupils' sustainable relationships, such as their behaviour towards one another. Both Nussbaum and Lögstrup are philosophers with an interest in ethics, Nussbaum on a more general level and Lögstrup on an individual level. Their theories thus complement each other.

Nussbaum (2010) argues that a country is obliged to produce democratic citizens, and to achieve this she suggests that subjects within the humanities and the arts are crucial at both primary and secondary school, and also at university. School and society need children who can negotiate well and who are able to have friends without making any of them their subject. Nussbaum refers to, among others, Jean-Jacques Rousseau who, according to her, made knowledge about basic human weakness central to his theories about teaching and learning. Children fight and tease each other, but a pupil who has developed a capacity for sympathy understands what their aggression does to another person. After insulting someone, this pupil experiences guilt about their aggression and wants to compensate the insulted person. Pupils recognize that other pupils are beings with their own rights.

Nussbaum (2010) identifies three pernicious structures when it comes to helping compassion and empathy to win over fear and hate. The first is where people behave badly when not held personally accountable. If pupils can act in a bad or mean way without having to take responsibility for their actions, it is more likely that this behaviour will increase. The same thing is likely if pupils have the opportunity to hide as part of a faceless mass. This is what often happens when a group of pupils gather and abuse an individual. In the second structure, people can continue to behave badly when nobody raises a critical voice. The pupils watching when another pupil is abused are often afraid and so do not dare to reprimand the abuser. The third structure is where people behave badly when those they have power over are dehumanized.

To work against these structures, Nussbaum (2010) argues that it is important that the school cultivates pupils by bringing them into contact with issues of gender, ethnicity, race and cross-cultural experiences. By using art and literature in teaching, pupils can learn to understand the achievements and sufferings of a culture different from their own. To sing, read and paint, to listen to music, poetry and novels, and to interpret a painting are examples of how to develop pupils' creativity and imagination. Nussbaum quotes John Dewey when she describes the importance of education for imagination:

> In a successful school, children will come to see that imagination is required to deal with anything that lies "beyond the scope of direct physical response." And this would include pretty much everything that matters: a conversation with a friend, a study of economic transactions, a scientific experiment. (p. 103)

Imagination is needed in different ways. In addition to being able to use imagination to see what is between the lines, it is needed for people to see their ideas as their own responsibility, and then they are more likely to see their deeds as their responsibility too (p. 54).

Furthermore, a trusting relationship is about responsibility. According to Lögstrup, we are forced to trust each other. In an interaction with someone you have a responsibility to encounter them in the way you believe they want to be received. In his book *The ethical demand* (1956/1992), he describes two opposite manifestations of life: the spontaneous and the confined. Trust and love are examples of spontaneous manifestations, and disagreement and suspicion are examples of confined manifestations.

The ethical demand is spontaneous; the common way to interact with someone is with trust. According to Lögstrup, trust is more fundamental than mistrust. As a consequence of this you always take a risk when you show trust and expose yourself to someone. That is why we react strongly when our trust is misused. Even worse is when our trust is not accepted. Conflicts that have nothing to do with right or wrong but come from our different natures and life-worlds become conflicts where moral and self-righteousness result in impossible accusations. In situations when no one has done anything wrong, the conflict has an emotional dimension, and from that comes a need to remake the conflict into something that is about having been treated in a wrong way. When the trust shown is not accepted, it motivates strong feelings. It is the emotional in the situation that makes one use moral reproach and accusations.

The ethical demand is also silent or unspoken. This implies that one always has to decide what the demand of the other person means. The way each of us chooses when we interact with someone will influence the other persons' life and might also affect how moral and ethical standpoints will be understood in the future (Lögstrup 1961). Another consequence of the ethical demand being quiet is that one has to avoid being too rigid and wanting to change another person, and also has to avoid being too elusive and escape trouble. The way one interacts with someone must be influenced by one's will to let the other person be master of their own life.

Lögstrup (1961) also writes that an individual can be held responsible for another person being insulted by a third person, at least in those cases where the individual is equipped with power but does not interrupt the violation. The reason why power is given to a human being is to prevent oppression. This is a responsibility for both teachers and pupils in a school.

All human beings have a responsibility for others. The pupils have a responsibility for one another and for the teachers, as well as the teachers having a responsibility for one another and all of their pupils. Notwithstanding, the ethical demand implies that teachers and pupils are given different responsibilities as a result of their different roles in school, and these different responsibilities interact with each other. What characterizes the ethical demand is not a demand for equal power but an understanding of the fact that teachers and pupils have different responsibilities and that these interact (Lögstrup 1983).

The Interviews

Interviews were carried out with 11 pupils in two Swedish compulsory schools: six pupils in grade 9 (age 15, four boys, two girls) and five pupils in grade 5 (age 11, two boys, three girls). The pupils were selected by their teachers. The teachers were asked to choose both boys and girls, and also pupils who differ when it comes to their way of handling social relationships with their friends in school. Each interview lasted between 15 and 30 minutes and was carried out in a small room at the pupils' school. For the interviews with the 11-year-olds, both the pupils and the parents had to give their consent to the pupils' participation. When it came to the 15-year-olds, the parents were informed and only the pupils had to give agree. The pupils, parents and teachers were informed that the purpose of the study was to find out more about pupils' views on how they improve their ethical literacy in school, and also how they think their school should work to create an attitude that supports sustainable relationships between the pupils. Table 1 lists the pupils who participated in the study.

Table 1 Participating pupils

Pupils—school A	**Grade**
Albert	9
Benjamin	9
Christopher	9
Daniela	9
Emma	9
Felix	9
Pupils—school B	
Gabriel	5
Hugo	5
Isabelle	5
Julia	5
Karen	5

The interviews were conducted in a semistructured way. Kvale (1997) describes this sort of interview as a conversation with structure and a certain purpose. The structure was formed by prepared areas of questions, which were the points of departure in the interviews.

Interpretation

In the work with the interviews, a hermeneutical interpretation has been achieved. A situation when a hermeneutical interpretation is used should, according to Gadamer (2005), be built on an expectation of openness about what is going to happen. To interpret means that one lets the world open up. In this work the ambition was to open up the pupils' worlds in school with regard to how they handle their relationships with one another. The 11 interviews are parts that form the whole of the material. They were transcribed and then the text was read several times. During the reading of the text, a pattern appeared. The parts of the pattern were mirrored against the whole material (Dahlberg et al. 2001; Ödman 2007) and against the theories of Nussbaum and Lögstrup described above. The researcher strives to be servant of the text, not the master (Palmer 1969). This implies that the ambition was to let the results reflect the pupils' views.

Findings

From the interviews a few aspects of ethical literacy appeared. The first part of the results reports on what the pupils view as reasons for them and others at their school abusing classmates even though they know it is not the right way to act. The second part reports on reasons why some pupils do not abuse their fellow beings. The third part describes how the pupils think their schools could prevent the unwanted behaviour.

Reasons Why Some Pupils Abuse Others

All of the interviewed pupils explained why they thought some people choose to abuse others in school. Some pupils admitted that they sometimes said and did bad things to their classmates. The reason, according to their stories, was mostly because someone else started a conflict by saying or doing something bad. When the pupils answered this question in a more general way, they gave three reasons: bad circumstances at home, a need to abuse others and feeling forced to abuse others.

Bad Circumstances at Home

Pupils can be nasty in school, and also in their spare time, as a result of difficult conditions at home. If one has problems at home, one might have a need to act out any bad feelings in school where one's parents cannot see. Two of the pupils explained it like this:

> Maybe they have a lot of quarrels in their homes and feel bad because of that and need to get their anger out on someone, even though that person has not done anything. (Karen, grade 5)
>
> It might depend on, well, that things are hard at home. And, well, they let it come out in school. (Albert, grade 9)

According to the pupils, their understanding of why some people behave badly towards others has its origin in discussions, both in school and at home. The explanation about bad circumstances at home seems to be something they had learnt from their parents.

A Need to Abuse Another

This explanation had its origin in stories from the interviewed pupils who reported that they often were, or had been, in conflicts with their classmates.

> you see that the other person becomes sad, but you just continue anyway. (Gabriel, grade 5)

Gabriel reported that he was aware of the other person's feelings but despite this there had been situations when he continued to hit the other person. He explained that he was full of adrenaline and could not stop himself. Later he experienced guilt and, after a lot of talking with his teacher and parents, and sometimes the headmaster, he claimed that he had learnt to control himself.

Felix explained that he tried to be nice to other people at school, but if they were not nice to him he would not put up with it—then he wanted to reciprocate.

> Well, how shall I explain, well if he is cold towards me, then I will be exactly like that towards him. I do not want to be kind to a person who does not show anything back. (Felix, grade 9)

Felix was not willing to set an example and ignore a bad comment: if he received one he needed to answer back.

Feeling Forced to Abuse Another

A common explanation as to why some abused others is that they felt forced to do so. If they did not take the initiative, they risked someone else doing so and them being the victim. These pupils thought that prevention was better than cure.

> They are afraid of being abused themselves. (Albert, grade 9)

Daniela had her own experiences of feeling forced to act in a bad and incorrect way:

> I knew it was wrong, but I did it anyway because I was so very lonely. (Daniela, grade 9)

She had no friends in her class, and to be spared from being on her own all the time in school she joined a small group of girls who demanded that she should do bad things if she wanted to be with them.

According to the 11 interviewed pupils, most bullies know what they are doing to the other person when they tease or abuse them, but there is a reason why they choose to abuse them. One might be that they are abused in their own homes. Another is that they are sometimes so angry that they need to lash out. It might also be that they are afraid of being abused or being alone. In spite of the fact that the 11 pupils understood that there is a reason why some pupils are mean towards others, they did not accept it as an excuse, just as an explanation.

Reasons Why Some Pupils Do Not Abuse Others

Most pupils do not bully or tease their classmates. They do not want to abuse anyone, and even if they become sad or angry because of another pupils' behaviour towards them, they do not retaliate. The interviewed pupils' statements about why they did not abuse their classmates are interpreted as expressions of ethical responsibility. The reluctance to be told off by teachers and parents is another reason some pupils stop themselves, as well as the intention to stand out as a good person.

Ethical Responsibility

According to the pupils, the difference between those who abused others and those who did not is a kind of ethical responsibility:

> I do not want to be mean and be given a rebuke. And I do not want to destroy for someone else. (Karen, grade 5)
>
> if you are being bullied then you will not be able to forget that for a long time, and it is quite serious if you fight or so. (Isabelle, grade 5)

As the quotes from Karen and Isabelle show, the pupils did not want to hurt someone else and accordingly they were willing to take responsibility when it came to how they chose to interact with other pupils, and in that way they supported sustainable relationships with their classmates.

Reluctance to Be Told Off

Some of the pupils, both those who often were or had been in conflict with other pupils in school, and those who avoided conflicts with their classmates, reported that to be told off by teachers and parents is not fun. They had learnt from such experiences and claimed that they were no longer in conflicts so often.

> I want to do well in school now. It is quite hard to get a telling-off and so (Gabriel, grade 5)
>
> I know what it feels like afterwards when you have done something, like "this I should really not have done" because I know what it feels like when you have to tell your parents and your teacher and you really do not want to. And then I always think before, that I do not want to come into this situation, it is not fun. (Emma, grade 9)

Both Emma and Gabriel reported that it was hard to disappoint your parents and teacher. The feeling of shame when they were told off prevented them from getting into conflict with others. They had both been involved in a large number of discussions when they had done something bad. This experience motivated them to stop themselves from getting into conflict with their friends.

The Teacher's Watching Eyes

Some pupils knew how to behave towards their classmates in the classroom, but during the break they might be the ones who abused others. According to the interviewed pupils, they behaved well in the classroom because they wanted to stand out as good people to their teachers. During the breaks there was no one to raise a critical voice or to watch them so it was easier to abuse a fellow pupil.

> You want to stand out as a good person, even though you may not be one. (Christopher, grade 9)
>
> During the lesson, in front of the teachers you want to stand out as good, but then you do it anyway, maybe because you want to do as your friends or because you are angry with that person. (Hugo, grade 5)

In these situations, the teacher's watching eyes were what prevented some pupils from abusing their classmates.

The pupils in this study gave three reasons why they did not abuse others. First, like many pupils, they were prepared to shoulder ethical responsibility and develop sustainable relationships.. Second, they did not want to risk being told off and disappoint either their parents or their teachers. Third, they wanted to stand out as good people in front of their teacher so behaved well in the classroom, even though they might abuse their classmates at other times.

How to Prevent Pupils from Abusing Others

Every day there are pupils who are abused by others in school. The pupils in this study were asked what teachers could do to prevent this. They believed that it was important to obey school rules, for which both pupils and teachers are responsible. They also suggested that all pupils needed to be made to understand the consequences of bad actions. Finally, they thought it was important for there to be a good atmosphere in the classroom.

Rules and Probation

Schools have rules, and it is important that when pupils do not follow them they are reprimanded. To improve engagement and understanding of school rules, one suggestion is that a school starts a council comprising only pupils who discuss the atmosphere among students and what to do when some fail to follow the rules.

> You could have something like the pupils' council where pupils from each class could talk about things that happen on the school. (Hugo, grade 5)

Another improvement would involve having more adults in the corridors and schoolyard during the breaks.

> There are not so many teachers in the corridors during the breaks, they see what happens during the lessons, but there are not so many teachers walking in the corridors just looking, so they do not notice small things and that is a big problem according to me. (Emma, grade 9)

As mentioned above, some pupils refrain from bad behaviour when a teacher might be watching, so if there were more teachers or other adults among the pupils between lessons, some of this would be prevented.

Improve Pupils' Insight Regarding Consequences

Those who bully others must learn what it feels like to be bullied, according to the interviewed pupils. Their insight regarding the consequences of their bad behaviour must improve in different ways, such as allowing the perpetrator to experience what it feels like to be abused.

> That you show the result of bullying, what might happen and maybe, in some way let the pupils feel, to show what it feels like to be abused. (Daniela, grade 9)

Parents and the school must take responsibility for teaching children the difference between right and wrong, making clear when a child has done something bad and when they have done something good.

> My parents have taught me how to behave and not to behave. If I do something wrong they tell me, and if I do something good they also tell me. They encourage me. It is more my home which has developed me to be who I am. (Benjamin, grade 9)

Parents and teachers can take deliberate action to prevent pupils from harming each other in various ways, in both the long and the short term. Pupils need to understand their responsibilities when they interact with other people. Several of the interviewed pupils suggested that children need to taught how to live with others in mutual respect.

Self-Confidence and Trust in Class

Another important condition for sustainable relationships in school, with the ambition that all pupils can feel safe, is that there is a good atmosphere in the classroom. One way to achieve this is for pupils to discuss things that happen between them. The teacher should be present, listening to what is said and helping to the move the conversation forward if needed, but not directing it.

> To just let the pupils in class sit down and discuss while the teacher goes round, and talks about what you think, that would help a lot. (Emma, grade 9).

To engage in activities other than schoolwork also improves the atmosphere and engagement in class. The pupils in the study reported that having a common experience from a variety of activities brings students together.

> "To give everyone the opportunity to feel happy in school you must do more activities together in the class" … "You will have memories from school, you want to have good memories, not just that you learnt mathematics in grade 8, but maybe that you went skiing with your class, 'those were lovely times'." (Felix, grade 9)

In order to feel self-confident and to have trust in each other in the class, pupils need to interact with each other in different ways. Talking about things that matter to them and having fun together makes it harder for them then to abuse a classmate and easier to understand and accept each other.

Discussion

This chapter highlights pupils' views about why some children abuse their classmates and others do not. I also want to shed light on what the interviewed pupils thought their school should do to prevent bad behaviour between pupils.

An overall objective of school is to develop pupils' ethical literacy—that is, to foster pupils' abilities to develop knowledge in different school subjects and values about, for example, equality and solidarity, and to give the pupils opportunities to learn how to describe and analyse ethics as a field of knowledge. Another objective, which is important to sustainable social relationships, is for all pupils to respect other peoples' intrinsic value. Despite this the interviewed pupils reported that violations to different degrees occurred almost every day in their schools.

Among the pupils who participated in the interviews, some reported that they occasionally abused a fellow pupil, and some of the participants explained that they did not want to do so. They had different experiences when it came to relationships with friends. All the same, the pupils' utterances corresponded with each other when they talked about why some abused other yet some chose not to. An ethical responsibility seems to be something that prevents one from abusing other people. This can be compared with what Lögstrup (1956/1992) calls the ethical demand. The pupils claimed that they did not want to hurt anyone or destroy anything for someone else. In that sense they were willing to take responsibility for how they treated their fellow beings and pupils. Furthermore, a willingness to stand out as a good person in front of the teacher and a reluctance to be told off by a teacher or parents are reasons to stop oneself from abusing others. In these situations the adults take a bigger responsibility than the pupils. This is also according to Lögstrup's ethical demand. All people have a responsibility when interacting with other, but adults have a greater responsibility. If pupils are to feel trust in their teacher, it is important that the teacher takes this responsibility (Lilja 2013). The teacher's watching eyes could be seen as part of this responsibility, but they cold also be seen as a risk, as Nussbaum (2010) shows, hence pupils may behave badly when not being personally accountable.

Previous research (e.g. Frisen et al. 2008; Varjas 2010; Thornberg 2010; Forsberg 2016) establishes that one reason to bully is to strive for power and status. The pupils in this study agreed. They said that people who are willing to hurt others are often afraid of being abused themselves, so to avoid this they are prepared to behave badly. Other reasons, according to the pupils, are bad circumstances at home and also a willingness to hurt. One of the pupils, Gabriel, reported that there had been occasions when he continued to hit even though he understood what he was doing to the other child. However, afterwards he felt guilt and his stomach hurt. This can be compared with Nussbaum (2010), who claims that children fight and tease each other, but when they develop a capacity for sympathy and an understanding about what their actions do to another person they learn how to behave and to take the responsibility that a sustainable relationship demands. Felix reported that when his efforts to be kind to some pupils in his class were rejected, he felt a need to retaliate. Lögstrup (1956/1992) writes about our strong reactions when our trust is misused. A situation like this can become a serious conflict—even though it has nothing to do with right or wrong from the beginning—because the emotions have been hurt.

Events that prevent bad behaviour between pupils include activities that encourage them to participate in decisions that concern themselves (Ahlström 2009). The pupils in this study highlighted this. Some of them suggested a council where they could discuss incidents and relevant rules. Another suggestion to prevent bad behaviour is that pupils have opportunities to discuss their relationships in class as well as matters that are close to them. This could be a way to support the vulnerable pupils, as Jennifer and Cowie (2012) suggest, but also to make it possible to adopt another person's perspective and imagine what it feels like to be abused. Imagination is needed in different ways, according to Nussbaum (2010), both to see what is between the lines and to learn that your own ideas are your responsibility. Sustainable relationships prevent people from abusing each other (Ahlström 2009). Trusting relationships are about responsibility (Lögstrup 1956/1992), and this applies to every interaction with another person. All pupils have a responsibility for others. The responsibility of the school is to create democratic citizens (Nussbaum 2010), and one way to reach this goal is to use the subjects in the humanities and the arts.

As Felix put it, it is important that you have fun with your classmates; then it is easier to be kind to each other. Maybe common experiences of an education with space for arts and the humanities can be a way both to have fun together, and to know how to argue and reason about what is good, bad, right and wrong, and how to practice these skills during the breaks with your classmates.

References

Ahlström, B. (2009). *Bullying and social objectives. A study of prerequisites in Swedish schools.* Umeå: Umeå Universitet. Social Studies.

Beem, L. A., Brugman, D., Host, K., & Tavecchio, L. W. C. (2004). Students' perception of school moral atmosphere: From moral culture to social competence. A generalizability study. *European Journal of Developmental Psychology, 1*(2), 171–192.

Bibou-Nakou, I., Tsiantis, J., Assimopoulos, H., Chatzilambou, P., & Giannakopoulou, D. (2012). School factors related to bullying: A qualitative study of early adolescent students. *Social Psychology of Education., 15*(2), 125–145.

Dahlberg, K., Drew, N., & Nyström, M. (2001). *Reflective lifeworld research.* Lund: Studentlitteratur.

Elvstrand, H. (2009). *Delaktighet i skolans vardagsarbete.* [Participation in the daily life in school]. Doctoral thesis, Linköping Studies in Behavioural Science, 144. Linköping: Linköpings Universitet.

Forsberg, C. (2016). *Teachers experiences of ethics in religious education.* Dordrecht: Springer.

Frisén, A., Holmqvist, K., & Oscarsson, D. (2008). 13-year-olds' perception of bullying: Definitions, reasons for victimisation and experience of adults' response. *Educational Studies, 34*(2), 105–117.

Gadamer, H. G. (2005). *Truth and method*. London: Continuum International Publishing Group.

Jennifer, D., & Cowie, H. (2012). Listening to children's voices: Moral emotional attributions in relation to primary school bullying. *Emotional and Behavioral Difficulties, 17*(3–4), 229–241.

Kvale, S. (1997). *Den kvalitativa forskningsintervjun* [The qualitative research interview]. Lund: Studentlitteratur.

Lilja, A. (2013). *Förtroendefulla relationer mellan lärare och elev* [Trusting relations between teacher and pupil]. Doctoral thesis, Studies in Educational Sciences, 338. Göteborg: Acta Universitatis Gothoburgensis.

Lilja, A. (in press). Teachers experiences of ethics in religious education. In O. Franck (Red), *Assessment in ethics education – A case of national tests in religious education*. Springer förlag 2016, Under utgivning.

Lögstrup, K. E. (1956/1992). *Det etiska kravet* [The ethical demand]. Göteborg: Daidalos.

Lögstrup, K. E. (1961). *Kunst og Etik* [Art and ethics]. Köbenhavn: Gyldendal.

Lögstrup, K. E. (1983). *System og Symbol. Essays* [Systems and symbols. Essays]. Köbenhavn: Gyldendal.

Nussbaum, M. C. (2010). *Not for profit. Why democracy needs the humanities*. Princeton/Oxford: Princeton University Press.

Ödman, P-J. (2007). *Tolkning, förståelse och vetande* [Interpretation, understanding and knowing]. Stockholm: Norstedts förlag.

Osbeck, C. (in press). Att resonera om vardagliga moraliska frågor – som utanförskap och kränkning [To reason about dayly moral questions – Like exclusion and violation]. In O. Franck, C. Osbeck, & K. von Brömssen (Red.), *Religioner, livsåskådningar och etik – för lärare årskurs 4–6* [Religions, outlooks on life and ethics – For teachers grade 4–6] (s. 247–260). Malmö: Gleerups.

Palmer, R. E. (1969). *Hermeneutics. Interpretation theory in Schleiermacher, Dilthey, Heidegger and Gadamer*. Evanston: Northwestern University Press.

Skolverket [Swedish National Agency for Education]. (2011a). *Läroplan för skolan, förskoleklassen och fritidshemmet* [Curriculum for the compulsory school, preschool class and the leisure-time centre]. Stockholm: Swedish National Agency for Education.

Skolverket [Swedish National Agency for Education]. (2011b). *Utvärdering av metoder mot mobbning*. [Evaluation of methods against bullying]. Stockholm: Swedish National Agency for Education.

Thornberg, R. (2010). Schoolchildren's social representations on bullying causes. *Psychology in Schools, 47*(4), 311–327.

Varjas, K., Meyers, J., Bellmoff, L., Lopp, E., Birckbichler, L., & Marshall, M. (2008). Missing voices: Fourth through eighth grade urban students' perceptions of bullying. *Journal of School Violence., 7*(4), 97–118.

Discourses of Available and Sustainable Lives: Ethical Literacy Offered to Tweens Through Fiction Reading

Christina Osbeck

Questions about sustainability are at the top of the societal agenda of our time. However, this does not mean that concerns about sustainability and sustainable societies are equally distributed among us. Our worries are to a large extent related to the contexts in which we live. Some people are participants in practices where these kinds of political issues are constantly present. Others suffer daily, practically, from the effects of pollution without being parts of the contexts where these issues are treated politically. Many of us contribute to the growth of the average ecological footprint without being sufficiently aware of what this means. Questions about how societies are organized when it comes to opportunities for experience, communication and learning are issues of the communicative economy of a society:

> An important question in a socio-cultural perspective [on learning] is how a society is organized in a communicative sense. What communicative experiences are various groups allowed to gain and what discursive environments do they have access to? What does a society's communicative economy look like and what are the opportunities to participate in qualified and development-oriented communicative activities distributed? (Säljö 2000, p. 248, my translation)

C. Osbeck (✉)
Department of Pedagogical, Curricular and Professional Studies,
University of Gothenburg, Gothenburg, Sweden

O. Franck, C. Osbeck (eds.), *Ethical Literacies and Education for Sustainable Development*, DOI 10.1007/978-3-319-49010-6_4

The communicative economy of a society is a question about societal sustainability.

Ecological issues should not be treated in isolation. On the contrary, they are related to how people live their everyday lives and what opportunities are available, what aspirations are present, and what forms of community and social cohesion these aspirations include. Sustainability is to a large extent about worldviews and ethics, which are also discussed more frequently in some contexts and less in others. Our repertoires of discursive practices—where we are active—are both different and of different scope, which means that the opportunities to imagine alternatives, and to review philosophies of life and habits, are unevenly distributed among us. Our ability to question if our available lives are also sustainable varies.

Fiction reading may be one way to compensate for limited experience, a way to widen our repertoire of discursive practices. It gives us the ability to empathize with the characters and in that way understand their experiences. If fiction is an opportunity for expanded repertoires of discursive practices, it depends among other things on whether it challenges existing repertoires. Do the stories offer visions of other available lives, and to what extent can these be understood as sustainable? The aim of this chapter is to present and discuss discourses of sustainable and available lives offered to tweens through fiction reading.

Theoretical Framework

Sustainability

As a concept, sustainability is multilayered (cf. e.g. Knutsson 2014). Often, ecological, economic and social aspects are stressed although they are also emphasized as being intertwined. In this chapter, overarching notions about a good life and good human relationships are foregrounded—that is, perceptions of social sustainability. These kinds of question can be interpreted as ethical because they concern what is right and wrong, good and bad, and what it might mean to be a good human being (e.g. Bexell and Grenholm 1997).

Martha Nussbaum's capability approach can be interpreted as an attempt to define what social sustainability and a sustainable life can mean. On the basis of Amartya Sen's theory Nussbaum suggests ten capabilities, centred on individual human dignity (e.g. 2001, p. 41ff.), that a state should guarantee each member of the state the opportunity to develop to a minimum level. This approach is in line with Nussbaum's earlier neo-Aristotelian writing where eudaimonia is the idea that should be given priority—that is, the full good human life in its many-sided wholeness (Nussbaum 1995). The capabilities are (1) to be able to live a human life of normal length; (2) to have good bodily health; (3) to have bodily integrity, such as being able to move freely and being secure against violent assault; (4) to be able to use senses, such as imagination and thought, shaped by, for example, education and experiences such as religious, literary and musical; (6) to reason practically, which includes being able to form a conception of the good and reflect critically—for example, about the planning of one's life; (7) to experience affiliation, and be able to live with and towards others with the social basis of self-respect, such as being treated as a dignified being whose worth is equal to that of others; (8) to live with concern for other species of animals and for nature; (9) to be able to play and enjoy recreational activities; and (10) to experience some control over one's environment, such as being able to participate in political choices and have property rights. Of these ten, Nussbaum states that affiliation and practical reason coordinate and permeate the others. Taken together, they can be interpreted as demonstrating, as in the case of eudaimonia, the importance of a holistic perspective on a human being and a full, good—sustainable—human life.

Ethical Literacy

Literacy is a concept that today is widely used in the human and social sciences even if it was previously primarily related to the linguistic sciences. The use of the term "literacy" in relation to ethics is not that frequent. A search of the EBSCO databases for peer-reviewed academic articles by journals yields about ten matches (May 2016), where the expression is used in various ways. Some researchers use the expression undefined (e.g. Whitmarsh 2009), others use it as moral awareness (e.g. Pless et al. 2011), while yet others link it to specific ethical traditions that are not always tightly linked to ethical literacy (Semetsky 2012). A common feature is that "literacy" refers to abilities of both an interpretative and an expressive, response-directed character. What this means varies. To show ethical literacy and be ethically literate is related to what in a certain practice is understood as being ethically literate (cf. Barton 2007, p. 185). Definitions are negotiated and shown in practice.

In this chapter, ethical literacy is understood in relation to sociocultural perspectives on moral development (Tappan 1992, 2006; Vestøl 2011; Thompson 2013). What is called "ethical literacy" here, and in the sociocultural tradition of Mark Tappan's "moral functioning", can be understood as depending on the repertoires of discourses that one has access to. Moreover, it is related to the practice where the individual is active, a context that affects what discourses and speech genres are privileged (Bakhtin 1986; Wertsch 1991). Sociocultural perspectives on moral development consider language, concepts and discourses to be cultural tools that are in use in moral functioning—that is, as "moral mediational means" (Tappan 2006). Varieties in discourses that one has access to, in what discursive practice one is active and the context of a certain situation are critical to understanding differences in ethical actions. The development of ethical literacy will in a concrete practice have a direction related to what in that context is understood as ethical conduct—that is, to be a competent actor in that practice (Säljö 2005, p. 140).

Expansion of Ethical Literacy Through Fiction Reading

According to Martha Nussbaum, fiction reading can be understood as a good substitute for experiences that we lack (e.g. Nussbaum 1995, 2008). Her hope in fiction reading has at least four aspects. First, we come into contact with destinies that we never would have the chance to be part of. Through sympathetic imagination we almost experience others' lives, explore them and try out other selves (cf. Conroy 1999; Lesnick 2006). Second, sympathetic imagination is cultivated when reading fiction, so that we become skilled in using that capacity in everyday life—that is, to be sensitive towards others' needs and to understand and empathize with them and their decisions. Third, fiction reading expands our imagination generally and offers a "knowledge of possibilities" (Nussbaum 2008, p. 145) so that we have the opportunity to see and be prepared for scenarios that might happen, and to have ideas about alternatives. Fourth, by showing scenarios that are not present in everyday life, fiction challenges our existential understanding so that new visions, hopes, possibilities and beliefs will take shape and in that sense have an impact on our ongoing re-creation of reality.

The possibilities that fiction reading offers can be understood as an expansion of repertoires of discourses, which are important in order to cultivate one's ethical literacy (Tappan and Packer 1991; Tappan 2010). Therefore it might be an advantage for ethics and values education to work with carefully chosen narratives (e.g. Conroy 1999).

The Context of Previous Research

A study of ethical literacy offered to tweens through fiction reading can be understood as a study in values education that includes both implicit and explicit forms (Thornberg 2008). Children's fiction reading can be both implicit values education—for example, offered discourses expressing different values—and explicit planned values education.

One risk with explicit values education is that it can become artificial. Studies have shown how lecturing in ethics seems to be less effective than discussions of moral issues (Cheung and Lee 2010), and that relevant everyday examples are of importance (e.g. Infinito 2003). This means that a central task is to find a balance between existing discourses and new, challenging ones that will contribute to the expansion of discursive repertoires and new possibilities.

Children's reading, and the socialization impact that such reading may have, has been paid attention to concerning both textbooks and fiction. Textbooks have been found to be the sources of both the stereotyping of others and mutual understanding (Opotow et al. 2005; Lee and Misco 2014) since they, for example, contribute to defining what a good citizen and a national identity might mean (Almonte 2003; Kwan-Choi Tse 2011).

Fiction reading as educational programmes has been found to primarily affect cognitive dimensions—for example, understandings of ethical concepts, and to a lesser degree affective and behavioural ones (Leming 2000; Berkowitz and Bier 2007). Cooperative work (Berkowitz and Bier 2007) and continuity in committing ethical discussions (Leming 2000) have also been shown to have a positive impact on the outcome.

Another focus is how pupils read fiction in ethical terms. Pupils can both stress and challenge boundaries between themselves and the characters of a text (Lesnick 2006). Distanced and judgemental readings (e.g. concerning some of the text characters) are forms of ethical commitment in reading where the boundaries are kept. By contrast, empathetic readings, in which one tries to understand from within, and affiliative readings, in which one shows group membership with characters, as well as opening readings, in which the story helps to share stories of one's own history, are forms of ethical commitment in reading where boundaries between oneself and the text are challenged. What such challenged boundaries mean in terms of the development of ethical literacy is in itself an important question. One can get the impression that new perspectives in that way are opened up, but research has also stressed the potential in keeping the distance in reading (Chinnery 2014). It is through realizing difference in a story, without reducing this difference to similarities in one's own life, that new perspectives can develop.

Possible explanations for why fiction may develop young people's ethical literacy are also a theme on this research agenda. A summary by Leming (2000) stresses four theoretical perspectives. First is a phenomenological one, emphasizing how fiction visualizes consequences of actions and how feelings are affected when given the possibility to experience lives never lived. Second is a cognitive moral developmental showing correlations between reading and moral capacity. Third, Paul C. Vitz's psychological research highlights that children's moral cognition might have the character of narratives, and that stories thereby could be advantageously used because they are close to children's ways of understanding what is moral. Fourthly, Albert Bandura's social cognitive learning theory—observational learning—is emphasized. There are similarities between social learning through observations and the process that the individual undergoes when reading fiction because one meets model acting.

A couple of Swedish studies have paid attention to values that are stressed or problematized in educational material for children related to sustainability and how identities are produced through these books. Eco-edutainment seems to produce environmentally aware and self-disciplined world citizens, a type of "ecological selves", who are expected to produce local eco-knowledge and eco-discipline in their families (Larsson 2012). Moreover, Ideland and Malmberg (2015) have shown how global threats and personal guilt are knitted together in texts and games in the education for sustainable development area and forced on children. However, education through fiction reading may have the potential to overcome this risk because of its ability to invite pupils to reflect on what a good and sustainable society might be and what challenges exist between "now" and "then", with a preserved respect for the integrity of the individual (Franck and Osbeck 2016).

The contribution of the current study to the research field is its focus on what tweens in Sweden read when they choose by themselves, and how this can be understood in terms of discourses of available and sustainable lives—that is, as contributions to their ethical literacy.

Materials and Methods

As a proxy variable for books frequently read by 10–12-year-olds, the most frequently borrowed books at the second largest library in Sweden, the library of the municipality of Gothenburg, were used (statistics from 2014). The library list is topped by two books from the same series, *Diary of a wimpy kid* by Jeff Kinney, which were borrowed 715 and 699 times, respectively. The next three books are *The silver boy* [*Silverpojken*] by Kristina Ohlsson, borrowed 634 times, *Room 213* [*Rum 213*] by Ingelin Angerborn, borrowed 631 times, and *An island in the sea* [*En ö i havet*] by Annika Thor, borrowed 580 times. As a reasonably large sample for this study, four books were examined. Out of the top two books of the same series, the first, *Greg Heffley's journal*, was selected and included in the sample. While the books about Greg are from the USA, the other three are Swedish. However, *An island in the sea* has been widely noticed internationally and in 1999 was awarded the German Jugendliteraturpreis. It has been translated into English with the title *A faraway island*.

Two research questions and two subquestions were applied to the material in order to respond to the aim:

1. What discourses of life—available lives—are offered to tweens through fiction reading?
   (a) How are human communities described and what is given value?
   (b) What can be interpreted as a competent actor in these communities?
2. To what extent can these discourses be understood as discourses of sustainable lives?

The first research questions (1a, 1b) were examined through close reading of the material in line with a hermeneutical tradition where a continuous shift between the interpretations of parts (as small as single sentences) and of wholes (as large as a work in its entirety) occurs. The reading resulted in emanating categories, and different ways of talking about and understanding human communities and values. These are presented here with examples from the books. However, the number of quotations has been restricted by the available space. The presented categories, while contextualized, can, in line with a definition where "discourse is a certain way of talking about and understanding the world" (Winther Jørgensen and Phillips 2000, p. 7), be understood as offered discourses of available lives. The second research question that guides the second-order analysis—that is, the analysis of the findings of the first research questions—is grounded in Nussbaum's capability approach as a way of grasping sustainability and sustainable lives.

The Contexts of the Four Books

Greg Heffley's journal is about a middle-school American boy. His experiences and everyday life in school, with friends and family, their ups and downs, is at the centre of this chronological diary.

In *Room 213*, Elvira goes to a summer camp where she meets new friends. The camp and close relations are foregrounded. A considerable amount of attention is paid to the heterosexual play between the girls and boys, but also to mysterious events related to a girl who died some years ago and happened to live in the same room as Elvira—Room 213.

Aladdin is the main character in *The silver boy*. He is from Turkey but lives in Sweden. His experiences and the course of life in his family—for example, how his parents get into financial troubles and are repeatedly robbed of food from their restaurant—are the hub of the story. Aladdin and his friends think this may be related to an old local event, to refugees in the neighbourhood or even to the presence of a ghost. Close relations are centred but placed in relation to shifting economical, societal and historical circumstances.

In *An island in the sea* an upper-class Jewish tween from Vienna, Steffi, ends up as a refugee in a small, remote and rather poor island outside Gothenburg, together with her sister, as a result of the Second World War. The story centres on experiences of being a newcomer, and the culture, class and relation conflicts that this can mean, but also on how, even in difficult situations, care can be shown. Here, too, the close relations are foregrounded but placed in shifting contexts so that the importance of their continuity is stressed.

Available Lives: Discourses of Communities and a Competent Actor

Greg Heffley's Journal

A first characteristic of the community described in this story is its hierarchical structure. The teachers create a hierarchy by grouping the pupils in relation to how skilled they are (p. 19). The hierarchy among the boys, which to some extent concerns popularity among girls, is related to clothes, wealth and good looks (p. 12). The school is a place where one tries to hide ones less advantageous features (p. 87) and one can gain points by exposing another person's (p. 53). This creates a context where Greg feels that he has to distance himself from a former friend: "I try to think much more about my image since I started in middle school. Then it is pretty worthless to have Rowley around" (p. 24).

Second, a gender stereotypical community is depicted. Very little is said about girls, and when they appear they are shown as bimbos or bitches, concretely as Greg's dreams about cheerleaders (p. 50), or complains about the boring and study-centred girls, telling the teacher (p. 100).

That people should have their minds set on doing as little as possible is a third characteristic. It is explicitly expressed as a strategy: "If there is anything that I have learnt from Rodrick [Greg's brother], then it is to set the bar really low, so that you in the end make people surprised even if you are nearly doing nothing at all" (p. 21).

A fourth characteristic is how one deludes and is deluded. It is acceptable to delude someone if you gain from it. It can be your friend, your parent, your teacher or your headmaster. Bad consequences for the other person are not a reason to avoid deluding them. There are examples of how one can delude one's friend, to put himself in danger or let him take the blame, as well as unfairly take credit yourself for their praiseworthy deeds (e.g. p. 188).

The last characteristic of the community is how success has intrinsic value. It is not related to the area in which you are successful or whether you deserve the success, although it seems to be related to the opportunity to exercise power, and to be seen and known (e.g. p. 8.)

To sum up, a competent actor in this community is described as a boy who likes a good laugh and who gladly laughs at the expense of someone else. He is a person who wants to control other people and make them do what he wants, with minimal effort. It is difficult to see any relationships that he would protect.

Room 213

A first characteristic of the community here is the presence of the supernatural. The book opens with a quotation that sets the atmosphere: "We are still seeing each other off and on, Bea, Meja and I. But we are never talking about *it*. It is a tacit understanding. As if it may not have happened if we do not talk about it. But deep down we know of course" (p. 5). In this way the story of what previously happened continues, although it cannot be talked about.

Second, the community is characterized as one where the opposite sex is shown attention as a possible partner. The most frequent example of a flirting method is teasing. To answer in a cool way that marks an (apparent) distance is another example (p. 156). The gender play is also used among the girls to elevate or lower each other. They build an internal hierarchy on how they appear in the eyes of the boys (e.g. p. 55).

Third, it is a context where friends tease each other and make practical jokes to such a degree that it is hard to say where the limit between a joke and naughtiness should be drawn. Coolness is marked by balancing on this limit and thereby risking crossing it. The first reaction when Elvira loses her ring is therefore that it is a joke, which is also the first thought that comes to the mind of the leader of the camp: "Jennifer [the leader] frowned. 'And you don't think that someone is joking with you?', she wondered. We shrugged. 'Though it's not exactly that funny…', Bea said" (p. 99).

Confidence in friends and care for friends is a fourth characteristic of the community. When all evidence points to Bea being the person who took the ring, Elvira still believes her: "In fact there was only one thing that didn't make sense. Bea herself. It just couldn't be her" (p. 71). Moreover, the story shows how friends support each other, defend each other, show loyalty and integrity, and not least laugh with each other.

A competent actor in this community is a girl with a well-developed social network and a high position among the boys. She controls others with her cool and quick tongue, but she also has the ability to feel and show affection. She holds the possibilities of the supernatural dimensions of reality open, and she protects relationships with close friends and potential boyfriends.

The Silver Boy

A first characteristic of the community in this story is how people care for each other—broadly. It concerns not only close friends but also parents, extended families and other people, such as priests and refugees. Friendship takes many forms, such as taking the initiative to help each other without being asked (p. 191 f.), or making a friend's bed in a nice way (p. 121). The concern for others is also shown in the family. Children can worry about the family's financial situation, and parents can try to keep these worries from their children (p. 19). To care for each other can take different forms—for example, anger can be a form of care (p. 33), and the care for the extended family can be shown as concern for grandma's life as a potential widow (p. 150). The care in the broader community can be shown as worries about the foreigner's possibility to get food, but also concrete actions to help (p. 144). Care as an interest in making use of each other's talents is also shown when one least deserves it: "None of us had seen the priest coming out of a door in a corner of the church [when, without permission, they started to play the organ and dance]. Thankfully he didn't look angry. On the contrary he smiled. 'How nice you play', he said to Aladdin. 'You should come here and play some day at service.' He looked at Billie. 'Then perhaps you could dance at the same time'" (p. 96).

Second, it is a community where economic, social and societal perspectives are present in the children's everyday lives. Children can be exposed to parents' conflicts, the tough economy and plans to immigrate (p. 37). The characters show n interest in history generally, and in their roots in particular, which extend beyond Sweden's borders (pp. 108, 238, 262). They worry about a polarized society, and they stress the importance of an active stand: "'Many people that live here have been very upset [about the refugee boat] and think that the people on the boat should go home again. And others, as you and I, bring them food' [Aladdin's daddy said]. Aladdin drew himself up. 'But in that case we definitely have to keep living here', he said angrily. 'Imagine if everybody that wants to share leaves!'"

Third, the way people act for each other is solution focused. A thread of the book is how Aladdin wants to help his parents with their financial problems and the thefts from their restaurant (p. 189).

A fourth characteristic is the belief that most things can be handled through wise actions and a conviction of never giving up: " 'You are so like your daddy', she said. 'To you nothing is impossible.' He blushed and shrugged. Some things were difficult and some things were easy. But impossible—no, almost nothing was impossible" (p. 259).

A fifth, supernatural, dimension is also present in this community. The book ends up stressing the possibility of the impossible: "One last time the boy with the short trousers followed them. [...] With quick steps he went over the snow and disappeared around the church. But any tracks of him where he had gone could not be seen" (p. 267).

What from this story can be grasped as notions of a competent actor is a responsible child who acts for his parents and friends, who wants to find solutions for difficulties and who is prepared to go far. Supernatural dimensions of reality are held open. Relationships with family and friends, but also with other people, in difficulty are protected.

An Island in the Sea

A first feature that characterizes the community in this story is the presence of care, but a care with many faces that can be hard to understand. When one, like Steffi, is used to oral and physical care, it can be hard to grasp, understand and appreciate forms of care that are much more practical, such as using extra cloth in a skirt when sewing (p. 169), or even disciplining and controlling (e.g. p. 73). Steffi realizes that there is a need for thankfulness, but since the care is hard to appreciate the thankfulness becomes false (pp. 34, 37). Care, for instance a teacher's care, can end up in exposed vulnerability where the risk of victimization will increase (p. 70). Moreover, care is often related to tenderness but could also be related to straightforwardness and hardness—that is, when someone that stands up for you even when it storms: " 'No one', Märta [the women who Steffi stays with] says. 'No one will come here and say something like that to my girl. He can be as fine as he likes. So there will be no apologizing.' " This shows that the described spontaneous thankfulness from Steffi is due to Märta's words rather than her act—Märta referred to her as "my girl" (p. 192).

A second characteristic is how cultural and class-related experiences can create misunderstandings. It can be hard to cope with differences in levels of standards of living that are too wide. How is it possible, for instance, to live in a place that seems to be the very definition of the end of the world (p. 18)? Furthermore, it may not be an advantage to be the best-dressed girl; quite the opposite (p. 68). And to move out of one's home in order to make room for vacationers may be hard to understand (p. 160). However, most difficult to cope with is perhaps a classmate's lack of understanding when together they look at a picture of Steffi's mother: "For a moment Steffi sees her mother with the eyes of Britta. The permed hair, painted lips, the coquettish fur boa around her neck. So different from the women on the island with their tightly set hair, faces without make-up and plain cotton dresses. She understands what Britta thinks about her mother: superficial and frivolous" (p. 87).

The community in school has a clear hierarchy that makes it difficult to be a newcomer, which is a third characteristic. A newcomer may destabilize the power structure, and some of the described bullying acts can be interpreted as being done to preserve the old balance. To coincidently succeed in changing the power structure may not be a victory to celebrate; quite the opposite (p. 91ff). Everything extraordinary can be a source of bullying, such as not being able to ride a bike (p. 67). To be a great imitator, and a bully, can be a talent that gives one credit in the hierarchy (pp. 71, 119). A strong hierarchy also means that one learns to despise people who are beneath oneself (p. 120), and that one has to socialize with people that take poor care of one (p. 79).

A fourth characteristic is that one has to accept the unacceptable. Several pieces of unwanted news are presented and they need to be accepted, such as the decision that one's parents are not allowed to become refugees in Sweden (p. 125), or that the doors to higher education are closed as a result of one's destiny, economic situation and the study tradition in the new context: " 'Things do not always turn out as one would wish', Uncle Evert says. 'We have to take life as it comes, and make the best of it' " (p. 143).

To manifest and ritualize community and shared happiness, despite the tragic historical situation and the modest circumstances at the island, is a fifth characteristic that can be grasped. The Easter bonfires and the celebration of the arrival of spring make the insight clear to Steffi: " 'On all islands', she thinks. 'On all islands there are people around the fire warming themselves. On all islands someone will ask if a child is cold. On all islands you can see the bonfire from the other islands.' The thought makes her happy" (p. 135). Budding trust and joy step by step come to Steffi: "She is not at the end of the world. She is on an island in the sea, but she is not alone" (p. 197).

So a competent actor in the context of this book might mean a responsible child who is prepared to do what it takes to survive and save oneself and one's sister. It is also a child who is aware of relational and societal difficulties, and knows that there are limits to what is possible. One has to try to make the best of it. Close relationships are most important but relationships with people in the neighbourhood are also vital. The interdependence of life is obvious.

The Offered Available Lives of the Four Books: A Brief Summary

The characteristics of the communities and the interpretations of notions of a competent actor that have been presented here can be understood as discourses of available lives. These available lives have recurring patterns.

There are hierarchical as well as gender stereotypical patterns. Supernatural dimensions are shining through but a strong care for others is also a repeated feature. The available lives of the four books vary. In *Greg Heffley's journal*, to delude and to be deluding is a strong theme. In *Room 213*, the importance of coolness is stressed but so is its downside. The belief in action-oriented solutions is emphasized in *The silver boy*, and how one is forced to accept the unacceptable is shown in *An island in the sea*. What could be understood as a competent actor in these books varies, among others along an egocentric–sociocentric dimension. *Greg Heffley's journal* is closest to the egocentric pole of the dimension while *The silver boy* and *An island in the sea* are closer to the sociocentric one.

Children who read these books already have discursive repertoires of life, about possible and desirable directions for the community and the individual. Reading gives them opportunities to expand their repertoires, and the characters of the books display other available lives. Whether these discourses expand the discursive repertoires of the lives of the children is an important question, even if it is a difficult one. We don't know the character of repertoires of discourses of life for specific children, and not that much either concerning children in general. Drawing on Sven Hartman's research on Swedish children's "thoughts about life" (e.g. 2000), a hypothesis would be that discourses of life including a societal dimension would have the potential to broaden children's perspectives because societal perspectives of life seem to be more frequent among older children than younger. That would mean that the four analysed books vary when it comes to a widening potential since their degree of societal perspective beyond personal and peer-group experiences differs. It is less in *Greg Heffley's journal* and *Room 213* and greater in *The silver boy* and *An island in the sea*.

Are the Offered Available Lives Also Sustainable Lives?

Expanded repertoires of available lives do not necessarily mean sustainable lives. The normative perspective draws on Nussbaum's capability approach, where individuals should have the opportunity to develop a minimum level of the stated ten capabilities in order to live a dignified life—that is, to have the freedom to choose what to be and do (2013, p. 29). A straightforward analysis would therefore involve examining to what extent the ten capabilities are present in the lives of the characters of the books. However, the meeting between the theoretical perspective and the inductive analyses of the books—focusing on the characteristics of the communities and notions of a competent actor—has not been straightforward. The analyses and their results tended to be detailed and hard to interpret fruitfully. Therefore, in addition to what can be understood as analyses of discourses of sustainable life on an individual level, analyses of discourses of sustainable life on a collective level have been conducted. Concretely, the latter form refers to analyses in search of discourses of communities where sustainable life on an individual level is protected (i.e. opportunities to develop capabilities). These are discourses where the focus goes beyond individual and group-centred interests towards an interest in fellow beings.

Systematic analyses of the four books have been conducted on the bases of sustainable lives both on an individual and on a collective level. Briefly, concerning the individual level, one could conclude that the main characters seldom live lives that are threatened or limited physically. The exception is in *An island on the sea*, where Steffi has been persecuted in Vienna and also lives a limited present life for socioeconomic reasons. Economic difficulty is also a thread in *The silver boy*. Moreover, it is striking that the capability to live with concern for other species and for nature is little represented in the books. It is only in the book about Steffi that such a concern is shown, in that case for dogs. In two of the books, *Room 213* and *The silver boy*, a concern for the dead through supernatural dimensions is noted instead.

Discourses of the good life as threatened seem to be connected to discourses of shielded opportunities for individual development and capabilities. Threats against available lives seem to open up discussions and actions regarding sustainable lives. In *Greg Heffley's journal* there is no threat against his available life and no discussion about a sustainable life. On the contrary, opportunities for dignified lives are threatened through Greg's actions, but these are not questioned. In *Room 213*, care for others is threatened through a peer culture of coolness, but actions to change this are not discussed. In *The silver boy*, threats against Aladdin's available life are present and actions are taken to help the parents. There is also an awareness of wider societal processes, resulting in concrete actions, such as giving food to people in need. In *An island in the sea*, the reason for Steffi and her sister coming to the island is that their available lives are not only threatened but also taken away from them. The actions taken for sustainable lives are taken both for themselves and for the sake of others (e.g. to help refugees). The four books differ concerning offered discourses of sustainable lives and, thereby, ethical literacy.

Ending Remarks

Perhaps one could say that *An island in the sea* takes a step further than providing discourses of sustainable life on a collective level by protecting individual opportunities for developing capabilities. A sustainable life—that is, here, dimensions of everyday life that sustain—is in this book also connected to realizing that there are opportunities available to those who can see them. However, this could also be understood as a capability that an individual has had the opportunity to develop, a capability that manifests itself through freedom to see and choose, which Nussbaum (2011) and Sen (2013), stress the importance of. Individual freedom and capabilities are from such a perspective resources that will solve emerging problems and "reshape the world" (p. 7).

However, neither individual nor collective capabilities develop automatically—they need to be sheltered. To contribute to sustainable lives and a sustainable society can largely be understood as an interest in what Säljö has described as the division of opportunities to qualified and development-oriented communicative activities—that is, the communicative economy of a society (2000, p. 248). In the absence of concrete experiences, fiction reading can be seen as one way of expanding one's discourses of life. If this expansion can contribute not only to discourses of available lives but also to discourses of sustainable lives, the selection of literature has shown in this study to be a critical factor, a factor for further consideration in values education—research as well as practice.

References

Almonte, S. A. (2003). National identity in moral education textbooks for high school students in the Philippines: A content analysis. *Asia Pacific Education Review, 4*(1), 19–26.

Bakhtin, M. (1986). *Speech genres and other late essays*. Austin: University of Texas Press.

Barton, D. (2007). *Literacy: An introduction to the ecology of written language*. Malden: Blackwell.

Berkowitz, M. W., & Bier, M. C. (2007). What works in character education. *Journal of Research in Character Education, 5*(1), 29–48.

Bexell, G., & Grenholm, C. (1997). *Teologisk etik: en introduktion* [Theological ethics: An introduction]. Stockholm: Verbum.

Cheung, C., & Lee, T. (2010). Contributions of moral education lectures and moral discussion in hong kong secondary schools. *Social Psychology of Education, 13*(4), 575–591.

Chinnery, A. (2014). On Timothy Findley's "The Wars" and classrooms as communities of remembrance. *Studies in Philosophy and Education, 33*(6), 587–595.

Conroy, J. C. (1999). Poetry and human growth. *Journal of Moral Education, 28*(4), 491–510.

Franck, O., & Osbeck, C. (2016). Challenging the concept of ethical literacy within Education for Sustainable Development (ESD): Storytelling as a method within sustainability didactics. *Education 3–13*, 1–10.

Hartman, S. G. (2000). Livstolkning hos barn och unga [Children's and young people's philosophy of life]. In E. Almén, R. Furenhed, S. G. Hartman, & B. Skogar (Eds.), *Livstolkning och värdegrund. Att undervisa om religion, livsfrågor och etik* (Skapande vetande, nr 37, s. 53–101). Linköping: Linköpings Universitet.

Ideland, M., & Malmberg, C. (2015). Governing 'eco-certified children' through pastoral power: Critical perspectives on education for sustainable development. *Environmental Education Research, 21*(2), 173–182.

Infinito, J. (2003). Jane Elliot meets Foucault: The formation of ethical identities in the classroom. *Journal of Moral Education, 32*(1), 67–76.

Knutsson, B. (2014). Utbildning för hållbar utveckling? Postpolitiska illusionsnummer och didaktiska alternativ [Education for sustainable development? Post-political illusory performances and didactical alternatives]. In O. Franck (Ed.), *A book of critique. Critical perspectives of policy documents, teacher education and school* (pp. 177–195). Studentlitteratur: Lund.

Kwan-Choi Tse, T. (2011). Creating good citizens in China: Comparing Grade 7–9 school textbooks, 1997–2005. *Journal of Moral Education, 40*(2), 161–180.

Larsson, B. (2012). The cosmopolitanization of childhood: Eco-knowledge in children's eco-edutainment books. *Young, 20*(2), 199–218.

Lee, L., & Misco, T. (2014). All for one or one for all: An analysis of the concepts of patriotism and others in multicultural Korea through elementary moral education textbooks. *Asia-Pacific Education Researcher, 23*(3), 727–734.

Leming, J. S. (2000). Tell me a story: An evaluation of a literature-based character education programme. *Journal of Moral Education, 29*(4), 413–427.

Lesnick, A. (2006). Forms of engagement: The ethical significance of literacy teaching. *Ethics & Education, 1*(1), 29–45.

Nussbaum, M. C. (1995). *Känslans skärpa, tankens inlevelse: essäer om etik och politik* [The sharpness of emotion, the insight of thought: Essays on ethics and politics]. Stockholm: Brutus Östlings bokförlag Symposion.

Nussbaum, M. C. (2008). Democratic citizenship and the narrative imagination. *Yearbook of the National Society for the Study of Education, 107*(1), 143–157.

Nussbaum, M.C. (2011). Creating capabilities: the human development approach. Cambridge, Mass.: Belknap Press of Harvard University Press.

Opotow, S., Gerson, J., & Woodside, S. (2005). From moral exclusion to moral inclusion: *Theory for Teaching Peace. Theory into Practice, 44*(4), 303–318.

Pless, N. M., Maak, T., & Stahl, G. K. (2011). Developing responsible global leaders through international service-learning programs: The Ulysses experience. *Academy of Management Learning & Education, 10*(2), 237–260. doi:10.5465/AMLE.2011.62798932.

Säljö, R. (2000). *Lärande i praktiken: ett sociokulturellt perspektiv* [Learning in practice: A socio-cultural perspective]. Stockholm: Prisma.

Säljö, R. (2005). *Lärande och kulturella redskap: Om lärprocesser och det kollektiva minnet* [Learning and cultural tools: About learning processes and the collective memory]. Stockholm: Nordstedts.

Semetsky, I. (2012). Living, learning, loving: Constructing a new ethics of integration in education. *Discourse: Studies in The Cultural Politics of Education, 33*(1), 47–59. doi:10.1080/01596306.2012.632163.

Sen, A. (2013). The ends and means of sustainability. *Journal of Human Development and Capabilities, 14*(1), 6–20.

Tappan, M. B. (1992). Texts and contexts: Language, culture, and the development of moral functioning. In L. T. Winegar & J. Valsiner (Red.), *Children's development within social context, vol 1, metatheory and theory* (s. 93–117). Hillsdale: Lawrence Erlbaum.

Tappan, M. (2006). Moral functioning as mediated action. *Journal of Moral Education, 35*(1), 1–18.

Tappan, M. B. (2010). Telling moral stories: From agency to authorship. *Human Development, 53*(2), 81–86.

Tappan, M. B., & Packer, M. J. (Eds.). (1991). *Narrative and storytelling: Implications for understanding moral development* (New directions for child development, No. 54). San Francisco: Jossey-Bass.

Thompson, P. (2013). 'It really hurts and it is bullying': Moral learning as political practice. *Journal of Moral Education, 42*(2), 224.

Thornberg, R. (2008). The lack of professional knowledge in values education. *Teaching & Teacher Education, 24*(7), 1791–1798.

Vestøl, J. M. (2011). Moral education and the role of cultural tools. *Journal of Moral Education, 40*(1), 37–50.

Wertsch, J. V. (1991). *Voices of the mind. A sociocultural approach to mediated action.* London/Sydney/Singapore: Harvester Wheatsheaf.

Whitmarsh, J. (2009). Developing ethical literacy in postgraduate research. *International Journal of Learning, 16*(3), 207–217.

Winther Jørgensen, M., & Phillips, L. (2000). *Diskursanalys som teori och metod* [Discourse analysis as theory and method]. Lund: Studentlitteratur.

Fiction at School for Educational Purposes: What Opportunities Are Students Given to Act as Moral Subjects?

Anna Lyngfelt

When Simone de Beauvoir claims (1982), "one is not born but, rather, becomes a woman", the same could be said about children in relation to childhood and children's literature. In particular, textbooks for children, including extracts from children's literature, tell us how children are supposed to become children by adopting educational aims. In fact, extracts from children's literature in textbooks could be said to evoke actions emanating from ideas in the texts of what children should be like.

In the school context, these actions are stressed by the functional, educational view of literature expressed in the syllabus for Swedish compulsory school, which states that students should get the opportunity "to read and analyse fictional texts and other texts for different purposes" (Skolverket 2011). This work should be intertwined with that to achieve the overall goals stated by the curriculum for Swedish compulsory school, stressing that students through their education should be able to express ethical standpoints, empathize with the situation of other people and distance themselves from the degrading treatment of people. Furthermore, language, learning and identity development are described as being closely associated with each other in the Swedish curriculum (ibid.).

A. Lyngfelt (✉)
Department of Pedagogical, Curricular and Professional Studies,
University of Gothenburg, Gothenburg, Sweden

O. Franck, C. Osbeck (eds.), *Ethical Literacies and Education for Sustainable Development*, DOI 10.1007/978-3-319-49010-6_5

The School Context as a Discourse of "Acts"

To discuss the school context, this context is here being viewed as a discourse of "acts" that have semantic meanings. To be more precise, this context will be narrowed down to the interplay between text content and actions evoked by *fictional texts in textbooks* written for students at compulsory school. However, this interplay is complex and not limited to explicit relationships. Rather, it is constituted by moral bonds within the school context and culturally related ideas regarding childhood and upbringing.

In this context, the theory of "acts", initially expressed by Edmund Husserl (1950), Maurice Merleau-Ponty (1945) and George Herbert Mead (1934), is relevant since it seeks to explain in what ways social agents constitute social reality through language. However, speech-act theory will here be limited to that elaborated by Kent Bach (inspired by Austin 1975). Bach's point is that utterances can be viewed as intentional action, in the sense that what one intends can contribute to what one does (Bach 1994). Together with his idea that discourse creates linguistic structures to construct the self, this concept might contribute to the understanding of *how* and *why* utterances in textbooks create possibilities and limitations for students to act as subjects within the school context. Consequently, identity is here regarded as a performative accomplishment, constituted by acts that one "does" (performs). Claiming that one is not born but rather becomes a child thus means appropriating and reinterpreting the doctrine of "constituting acts" because being a child is in no way a stable identity or locus of agency from which various acts proceed—rather, it is an identity tenuously constituted in time, an identity instituted through a repetition of acts.

Children's Literature as a Repetition of Acts

From an educational or moral point of view, what children need has often been studied. In fact, the history of children's literature could be said to be a repetition of acts, maintained by authors creating children's literature and writers focusing on children's literature. It is worth noticing that even if writers of literature do not intend to be advisory, but instead pursue imaginative freedom, the results can be the opposite.

Of course, this could also be said about researchers. For instance, while criticizing the unimaginative, lecturing content of older textbooks, Paul Hazard lectures on what children are and what they need:

> It is true that they lure us away from the feast of ideas, taking no pleasure there themselves. They place small value on the abstractions that are so useful to our grown-up pastimes. No doubt it sometimes happens that the stories most laden with meaning seem easy for them to take hold of, as though they had already lived several lives, the confused memory of it survives; or as if they had foreknowledge of their own completion; or as if intuition accomplished its miracle in them and allowed them to reach their goal while sparing them the journey to it. But these are only exceptional rays of light. Let us not exaggerate, let us not grant them every quality, let us admit that they have no skill in handling ideas. What they have is enough for them. (Hazard 1947, p. 166)

Authors, too, have explicitly expressed opinions about what children need. For instance, Zacharias Topelius, as early as 1865, stressed the importance of texts not patronizing children, a standpoint which may be compared with ideas expressed a hundred years later by another author writing for children: Lennart Hellsing. In 1963, Hellsing wrote about how important it is that texts for young readers (1) interest them; (2) have sufficient aesthetic merits; and (3), if they treat ethical matters, keep up to "the morals or tendency that we are likely to accept" (Hellsing 1963, p. 52, my translation).

That a distinction is possible to make, separating children's literature from that for grown-ups, Örjan Lindberger discusses in his research (1998). Demonstrating examples of adaptation of literature written for adults into children's literature, he points out what historically has been regarded suitable for children. Lindberger shows that there are several characteristics of literature that have been adapted: (1) the content of the transformed literature concentrates on adventures or stories about children, families or animals; (2) the story itself is concrete and not too slow (when it comes to action); and (3) children can relate to the stories. He also writes that complicated language should be avoided in adapted literature, as well as (hidden) symbolism that could be said to prevent young readers from understanding the meaning of the text.

In her work, Jacqueline Rose is aware of the didactical implications of conceptions of children's literature (1992). In *The Case of Peter Pan or the Impossibility of Children's Fiction*, she foregrounds the complex potential meanings of children's literature by asking what it is that adults (through literature) want or demand of children. She regards children's literature as something that "seems to operate according to a regime of attraction which draws the child straight into the path of identification—with the intimacy of the story-telling itself, or with the characters in whom the child recognises himself or herself on the page" (p. 140). Here this is interesting, since Rose explains this process as an activity of oppression of children by adults. As she puts it, every reader (adult as well as child) has to take up a position of identity in language by recognizing themselves in the first-person pronoun and "cohere themselves to the accepted register of words and signs". Problems arise, she says, when this necessity of recognition shifts into "something more like a command" (p. 141).

Rose's research, as well as the theories on adapted literature, problematize the fact that even if adults have the intention not to (consciously) have an impact on literature for children, adults constantly keep filling children's literature with a content that reflects their own ideas of what is suitable for young people to read. Whether it be moral proclamations—typical of children's literature in the past—or something else, children's literature is likely to express moral standards that are desirable within the society in which the texts are being produced. In that sense, both the young reader and the main character in children's books could be said to be objects rather than subjects of constitutive acts, contributing to the performativity of childhood and the creation of what they (themselves) should be like.

Studying Extracts from Children's Literature

To be able to analyse messages conveyed by contemporary children's literature, and discuss the performativity of children's literature in the textbook context, the focus here will be on fiction in textbooks for pupils aged 10–15. Three texts from three textbooks are discussed, as well as exercises relating to the texts.

The following books have been analysed: Alvåker and Boglind's *Bokskåpet* (1994 [*The bookcase*]), Falkenland's *All världens berättelser och du* (1998 [*All stories of the world and you*]) and Sjöbeck and Holmström's *Svenska nu* (1999 [*Swedish now*]). In all three, literature is a major part of the content. However, these books have also been chosen because they differ from each other. *Bokskåpet* is similar to a textbook anthology but with exercises linked to some of the texts. On the one hand, these exercises aim to develop the thoughts evoked by the texts (e.g. "Write about a time when you experienced something very sad."—all translations related to the textbooks are my own). On the other hand, they aim to help the young reader learn about narrative technique.

On the back cover of *All världens berättelser och du*, the book is referred to as a "study anthology". Every text is presented with an introduction and followed by exercises. The exercises consist of tasks that encourage students to dramatize the content of the stories or to talk/write about them. (The aim is, as Falkenland puts it, to get pupils to reflect on what they have read and take a stand where they compare their reading experience with practical experience from real life; see Falkenland's preface, p. 5.)

Svenska nu uses an approach characterized by an integration of all kinds of schoolwork included in the subject of Swedish. Extracts from children's literature are combined with exercises aimed at speaking and writing practice, and even questions about grammar. The book also aims to cover the whole subject, as well as the overlaps between Swedish and subjects such as history, religion and civics. Fiction, here, can be said to be the starting point for broader studies than are traditionally included in the subject of Swedish.

In this chapter, three texts from the aforementioned textbooks are read more closely. They are all written by contemporary Swedish authors writing for children. The texts under study are Lars Hesslind's "Använd språket, grabben" ("Use your language, young man" from *Bokskåpet*, pp. 148–155), Mecka Lind's "När mamma rymde" ("When mum ran away" from *All världens berättelser och du*, pp. 106–112) and Shanti Holmström's "Ur Mitt okända hemland" ("From my unknown home country" from *Svenska nu*, pp. 8–13). One of these stories was written by a man (Hesslind) and the other two women (Lind and Holmström), which could be said to be a typical gender representation of publications on children's literature in Sweden. (Most textbooks include extracts from children's literature.) The stories were chosen because they represent three different perspectives on social issues: "Use your language, young man" is interesting from a socioeconomic standpoint, while gender studies are relevant to "When mum ran away". Questions about ethnic belonging and identity could be said to be relevant after reading "From my unknown home country". (From now on I refer to the titles translated into English to make it easier for the reader to follow.)

In all three stories the reader meets a main character who in the first person tells of something that has happened in the past, and by doing so they both (re-)present the book content and evoke action denoting performativity. Though two of the stories are extracts from longer stories

("From my unknown home country" and "When mum ran away"), this will not be taken into account; after having read the whole of the stories, it is clear to me that the extracts are typical of the stories as a whole. (Both are from the beginning of the stories.)

Performing the Child Through Extracts in Textbooks

"Use your language, young man" is set in the 1940s and is about a working-class boy in a Swedish town who hits his long-time antagonist with an icy snowball. The boy who was hit swears revenge. However, the revenge never happens because the main character has learnt how to rattle off the (classical) Greek alphabet. "You can't beat someone like you", the antagonist argues, "you're so crazy I wouldn't even dirty my hands on you" (*En sån som dej kan man ju inte slå. Du är ju så jävla flängd i huvudet att man inte kan skita ner händerna på dej*, p. 154). At the end the main character tells the reader that he has learnt from this that words are the strongest weapons in the world.

The story is written in the first person, yet it includes comments that are more likely to be spoken by a grown-up than by a child, such as "From the structure of our language you may judge yourself, how far we were from the bookshelves where we grew up" (*Av språkets struktur kan ni själva räkna ut hur långt det var till bokhyllorna där vi växte upp*, p. 149). The text being written in the first person, from a young boy's perspective, the presence of irony is remarkable ("Life was damn good", *Livet var skitkul*, p. 152). The choice of words, stylistically, also contributes to the alteration of the social agent's position, from being in the position of a subject, retelling in the first person, to that of an object. For instance, at the end, it is said about the antagonist that "The expression on his face slowly changed, from the one of a scornful avenger to one of a doubter" (*Uttrycket i hans ansikte förändrades sakta från den hånfulle hämnarens till tvivlarens*, p. 154). Together with the content of the story, the stylistic level strengthens the positioning of the adult as a subject. This happens, for instance, when the boy meets an artist, who tells him that his canvas invites him to dance with "the anxieties of creativity" (*där skapandets ångestar bjöd upp till första dansen*, p. 151).

"When mum ran away" is also written in the first person. The narrator explicitly informs the reader that the main character, a girl called Ollie, is 11 years old. The story is retold from her point of view and is about an ordinary afternoon ending in chaos. It takes place in a flat where a (single) mother lives with her three children: Ollie, Jesper (who is younger than Ollie) and Tessa (who is about 15). They are all quarrelling with each other in the kitchen while their mother is frying fish fingers. They remark on the food that is being prepared: “Actually, you do get paid for us, mum” (*Du får faktiskt betalt för oss, morsan*), Tessa remarks (p. 108). “First, you get the child allowance, then you get money from dad, too” (*Först får du barnbidrag och sen betalar pappa för dig också*), she goes on (p. 108).

This and other comments finally make the mother feel that she has had enough—at least the reader is supposed to think so when the children find themselves left on their own the following morning when they wake up. Their mother has left a note explaining that she cannot endure anymore and that “she needs to rest and think” (*Måste få vila och tänka efter*, p. 111).

At the beginning it is relevant to argue that the story—literally—is being told not only from the main character’s point of view but also from the other children’s perspectives; the reader receives information about what is going on through the dialogue. The focus of the story gradually changes, though, from that of a social agent (the main character) being a subject to the being an object (when the mother’s perspective takes over). This development is not (mainly) expressed in words but through the mother’s body language. (You can read that the mum first bangs a saucepan on the stove. Then she bangs a frying pan so hard that she makes the fish fingers jump like dolphins. Later she rushes into the bedroom and locks the door.)

Yet, stylistically, nothing (verbally) reveals the transformation from the main character’s position of a subject to the position of an object (related to the mother’s position as a subject). So if the main character’s “object position” in “Use your language, young man” is not very well hidden (in terms of expressions and vocabulary with references to a world far from what could be said to be ordinary childhood experience), “When mum ran away” verbally sticks to the “subject position” while letting the children dominate as social agents.

The third text, “From my unknown home country”, is similar to a diary. The reader gets to know a teenage girl called Shanti, who describes what it is like to be adopted and look different from many other Swedish young people. Her thoughts keep wandering about in the text, from her biological parents and the parents she is living with now, to the experience of having found herself close to a neo-Nazi demonstration when visiting Stockholm.

Since the text exposes Shanti's thoughts in a convincingly realistic and personal way, and the language is not very complicated, the text can be said to stick to the main character's subject position. Yet if the content of the story is being analysed, it has—just like the other—a sense of morals that transfers the main character to an object position. When Shanti talks about the demonstration, for instance, she tells the reader that the demonstrators acted without knowing what they were doing (*De visste ju egentligen inte vad de gjorde*, p. 9). She is also remarkably sensible when it comes to people who have hurt her. For instance, she tells the reader that she understands a tormenting skinhead who has been bullying her, and suggests that they could have talked to each other because they "might have been able to help each other" (*hade kanske kunnat hjälpa varandra*, p. 11). Throughout the story, Shanti keeps her feelings under control and tries to understand the world around her, and she expresses a forgiving attitude even when people are being mean to her.

Extracts with Exercises in the Textbooks

No specific exercise is linked to "Use your language, young man", while several tasks are set after "From my unknown home country". The text is part of a chapter entitled "Read and write about identity" (*Läs och skriv om identitet*, pp. 7–21), making it clear, literally, that developmental psychology could be said to have an impact on the Swedish school system.

First the students are asked to retell the content of "From my unknown home country", then they are asked to describe themselves using nouns and adjectives. After this they are supposed to make a timeline describing their lives so far, and discuss questions such as "What possibilities does one have to control one's own life?" (*Vilka möjligheter har man att styra sitt liv?*), "Does it make any difference what you do?" (*Spelar det någon roll vad du gör?*) and "Should you plan your life and already be working on your future—for example, by studying?" (*Ska man planera livet och redan nu lägga grunden till framtiden genom att till exempel studera?*, p. 17). The students are also asked to write a story about their future, where they fantasize about what it is going to be like. At the end of the chapter in the textbook, they are asked to "investigate different lifestyles and different things people believe in" (*undersöka olika livsstilar och olika saker som människor tror på*, p. 19). Finally they are asked to make a self-evaluation, working with questions such as "Why may it be a good thing to think of what shapes one's identity?" (*Varför kan det vara bra attfundera över vad som format identiteten?*) and "Are you pleased with your own work?" (*Är du nöjd med ditt eget arbete?*, p. 21).

Here these exercises are described to focus on how the interpretation of children's literature is directed in a textbook context; even if subject positions can be said to be present in the literary texts, they tend to disappear when the work with the texts begins. This is especially true in questions such as "Should you plan your life and already be working on your future—for example, by studying?" and "Why may it be a good thing to think of what shapes one's identity?" (exercises set after "From my unknown home country"). Yet it is not as simple as saying that a text such as the one about Shanti includes the positioning of a subject, while the textbook context does not offer opportunities to analyse it from a young person's subject position. It is, for instance, possible to argue that "From my unknown home country" has a form that gives the reader an impression of a subject positioning herself (the "diary"), while the content (or rather the message conveyed) indicates the positioning of an object. Also, studying the questions one could argue that the opportunity to act as social agents *is* possible within the framework of doing the exercises—at least to some extent the pupils can leave their marks on the answers.

Similar thoughts are evoked by "When mum ran away" and the questions related to this text. Below the heading after "When mum ran away", "The story and you" (*Berättelsen och du*, p. 113), the following questions are found: "Whose 'fault' do you think it is when there is a quarrel in Ollie's family?" (*Vems "fel" tycker du att det är att det bråkas i Ollies familj?*), "Was it 'right' that the mum ran away, do you think?" (*Var det "rätt" att mamman rymde, tycker du?*), "Why did she run away, do you think?" (*Varför rymde hon, tror du?*), "Is a parent allowed to run away?" (*Får en förälder rymma?*) and "What could she have done instead?" (*Vad kunde hon ha gjort istället?*). The questions offer the opportunity for the students to position themselves as subjects in the answers because they are open and they invite answers from a personal point of view. Yet the educational purposes of the exercises and the textbook are likely to have an impact on the students when answering the questions.

Performativity Through Extracts

In fact the text content could be said to signal performativity in the sense that children denote modes of presenting the content (from an object position), while adults denote modes of evoking action (from a subject position). In "Use your language, young man", for instance, this can be seen when the young boy says that he realizes that language is the strongest weapon in the world, and in "From my unknown home country", the main character acts and draws conclusions as any adult positioning themselves as a subject would (which could be regarded as the ultimate act of positioning the main character as an object). Also, remarks about the situations consist of expressions that deviate stylistically from what might be supposed to be typical of children's language and references (especially in "Use your language, young man").

On the other hand, in all three stories, children can be said to denote modes of subjects positioning themselves as well: all of them are written in the first person, with young people as (fictional) storytellers. All of them also depict situations where the main character initially positions themselves as a subject: in a conflict increasing because of a snowball (in "Use your language, young man"), during a rowdy afternoon at home (in "When mum ran away") and in an ordinary situation such as eating at school (in "From my unknown home country").

However, in all of the stories an educational purpose can be discerned, as well as in the exercises linked to them. Perhaps educating the pupils with the aim of telling them that learning language is important (and the right use of it even more important—see "Use your language, young man"), that being nice to your mother is good (and to your brothers and sisters—see "When mum ran away") and that trying to understand other people is desirable (even if they bully you because of your skin colour—see "From my unknown home country ") offers the most striking examples of children having altered their positions of being subjects to being objects.

Of course, this might be regarded as the result of the fact that texts written for children also need to suit adults—at least to some extent (to be sold and, in this context, accepted by teachers). As pointed out, another explanation is the history of children's literature as morally educational. Worth stressing here is that the tradition of morally educating children through children's books risk taking place and growing stronger through the support of the educational context at school (including exercises). In fact, one could argue that the school context opens up for the *double adaptation* of literature in textbooks, partly as a result of the idea of children's (supposed) needs in general and partly because of the moral standards maintained by the school discourse.

Undoubtedly this requires a discussion about what texts are being used at school and why; if discourse creates linguistic structures to construct the self, the possibilities and limitations created by the choice of texts is crucial (Bach 1994). In addition to this, the students' tasks related to the text extracts make it clear that it is important to discuss in what ways texts used in school make it possible for students to act as subjects within the school context. Furthermore, if the goal in the curriculum of students being able to express ethical standpoints is to include their own ethical standpoints, they need opportunities to act as moral subjects while reading and discussing fiction at school.

Fiction at School and the Opportunity for Students to Act as Moral Subjects

To enable sales, are textbooks made to suit as many readers as possible with as little confrontation as possible, as the Swedish author Sven Wernström puts it? According to him, texts in textbooks are predictable and (therefore) uninteresting (1991). What students need, he points out, is fiction that opens up the unpredictable in life, showing the irrationality of peoples' behaviour and their (more or less) repressed motivations for their choices in life.

Undoubtedly students need fiction to help them to grasp complexity if they are to discuss interpersonal complex situations at school. Wernström might well be right when he says: "To be able to act as a human being there is only one user manual, and that is fiction" (p. 167).

References

Alvåker, L., & Boglind, A. (1994). *Bokskåpet* [The bookcase]. Malmö: Gleerups.

Austin, J. L. (1975). *How to do things with words.* Oxford: Oxford University Press.

Bach, K. (1994). *Thought and reference.* New York: Oxford University Press.

de Beauvoir, S. (1982). *The second sex.* London: Harmondsworth.

Falkenland, R. (1998). *All världens berättelser och du* [All stories of the world and you]. Stockholm: Natur och Kultur.

Hazard, P. (1947). *Books, children and men.* Boston: Hornbuck.

Hellsing, L. (1963 [1999]). *Tankar om barnlitteraturen* [Thoughts on children's literature]. Stockholm: Rabén & Sjögren.

Hussserl, E. (1950). *Husserliana. Gesammelte Werke.* Haag: Nijholt.

Lindberger, Ö. (1998). Barnböcker och vuxenlitteratur. In K. Hallberg (Ed.), *Läs mig – sluka mig! En bok om barnböcker* [Read me – Swallow me! A book on children's literature] (pp. 19–32). Stockholm: Natur och Kultur.

Mead, G. H. (1934). *Mind, self, and society: From the perspective of a social behaviorist.* Chicago: University of Chicago Press.

Merleau-Ponty, M. (1945). *Phénoménologie de la Perception.* Paris: Gallimard.

Rose, J. (1992). *The case of Peter Pan or the impossibility of children's fiction.* London: Macmillan.

Sjöbeck, K., & Holmström, F. (1999). *Svenska nu* [Swedish now]. Stockholm: Liber.

Skolverket. (2011). *Läroplan för grundskolan, förskoleklassen och fritidshemmet 2011* [Curriculum for the compulsory school, preschool class and the recreation centre 2011]. Stockholm: Skolverket.

Topelius, Z. (1865). *Läsning för barn* [Reading for children]. Stockholm: Bonniers.

Wernström, S. (1991). Ös böcker över barnen! In T. Karnstedt (Ed.), *Läsfeber. Om läsandets lust och vånda* [Reading fever. On reading for pleasure and Anguish] (pp. 163–168). Stockholm: Litteraturfrämjandet.

An Aesthetic and Ethical Perspective on Art-Based Environmental Education and Sustainability from a Phenomenological Viewpoint

Margaretha Häggström

Introduction

Experience and personal involvement are crucial to education and learning processes, in accordance with a pragmatist tradition (Dewey 2009). They are also central to phenomenological studies. Phenomenological learning emphasizes the process of meaning-making through an individual's self-experiences (Selvi 2009). Pupils' self-experiences include relationships with others, and Selvi (2009) claims that pupils comprehend the core of their experiences in relation to learning topics. She suggests that the aim of phenomenology is to search for the meaning of the world and objects. Dewey (1958) argues that there is a relationship between perception and the environment of the individual, and that perception is a reaction to the unpredictable features of nature. In light of the philosophies of pragmatism and phenomenology, one could argue that self-experience and relationships with individuals' environment is crucial to their understanding of the world. One starting point in order to create opportunities

M. Häggström (✉)
Department of Pedagogical, Curricular and Professional Studies,
University of Gothenburg, Gothenburg, Sweden

O. Franck, C. Osbeck (eds.), *Ethical Literacies and Education for Sustainable Development*, DOI 10.1007/978-3-319-49010-6_6

for self-experience and relationships with the natural environment in the school context could be to implement art-based environmental education.

Here I discuss school pupils' life-worlds and art-based environmental education from a phenomenological perspective. The chapter is structured in three parts. First, I describe the philosophical framework used to understand pupils' life-worlds, drawing on the work of Husserl, Heidegger, Merleau-Ponty, Schütz and Bengtsson. The notions of plant blindness and otherness are also discussed. Second, I present art-based environmental education as a teaching and learning tool to embrace bodily knowledge and to meet the demands of sustainable education, according to the United Nations Educational, Scientific and Cultural Organization's (UNESCO's) policy documents and Swedish curricula.

Third, I outline a general idea of sustainability and education in relation to the work already mentioned. Concluding thoughts in relation to teaching and learning from a phenomenological perspective end the chapter.

The Phenomenological Perspective and the Life-World

A phenomenological approach seeks to understand people's experiences, attitudes and perceptions of different phenomena. The fundamental structure of an experience is its intentionality (Merleau-Ponty 1995). This means the way the experience is directed through its meaning towards a certain object. Merleau-Ponty (1995) claims that we always turn our attention to something and that intentionality exists in our consciousness, in our body. While we live through or perform an experience, we simultaneously experience an awareness which makes the experience conscious. How we turn our attention to something may have an impact on how we perceive and understand it, and how we speak of it (Segolsson 2011).

According to phenomenology, knowledge is embodied and therefore bodily experience is essential to learning processes (ibid.). A phenomenological perspective may provide a deep understanding of a phenomenon in a new way. This means that the phenomenon itself can be well known at first, but not from a certain viewpoint.

The phenomenological life-world perspective originates from the philosophical movement and the works of Husserl (1989), Heidegger (1993), Merleau-Ponty (1995) and Schütz (2002), and the specific strand I am using here has been developed at the University of Gothenburg, mainly by Bengtsson (1988), Claesson (2009) and Berndtsson (2001).

The focus in life-world research is people's experience in relation to a specific phenomenon, and how they understand and construe their experiences. Thus it is not a matter of *what* but of *how*, or rather *how* people experience the *what*—the ways we experience things. Furthermore, this perspective tries to understand how an individual creates meaning from their experiences. One assumption is that we experience the world through our senses and bodies, and through experiences, self-experienced or mediated by others.

Our existence in the world is corporeal and we cannot separate ourselves from our body; it is our access to the world. According to Merleau-Ponty (1995), the life-world concept is a body and subject theory, and the lived body is the subject of all experiences.

The life-world is where we live our everyday life. We take the world for granted. Thus we can assume that the world is still there when we wake up in the morning (Claesson 2009). The everyday life-world is, according to Schütz (2002), superior to all other temporal worlds, such as the dream-world, theatre and art-world, or children's play-worlds, which are all seen as modifications of the everyday life-worlds.

Being-in-the-World

Through the concept of being-in-the-world, Heidegger (1993) postulates that we are in the world as historical individuals—that is, we carry the past and simultaneously address ourselves to the future, but we make our decisions in the present. This three-fold structure is composed of how and who we were, are and will become. Our struggle for the future makes the life-world changeable and dynamic. Being-in-the-world implies a sense of time, and the awareness of temporality establishes our relationship with the world. According to this perspective, we are capable of taking care of the world, although it is our future and the next generations' future. If we did not have a sense of time, we might not be able to understand how we affect the world. In other words, if we could not look back at the past but just stay in the present, we would not know anything about the continuity and we would not know anything about tomorrow. As Griffiths (1999, p. 14) puts it, "To live in a synthetic ever-present present is to live not in the fullness but in the emptiness of time." But this is still only from a human perspective; time exists everywhere in nature. It is explicitly notable during the four seasons, especially while following a plant from spring to winter. Every month, week and day has an agenda for the plant, or rather the plant has its own agenda for each day.

We are in the world, not only as historical individuals but as social and cultural individuals too. Therefore the life-world can never be an individual project. We share the world and we are part of each other's worlds. Consequently we are able to understand each other and our worlds. This understanding is a necessity for all of our internal and external knowledge (Nielsen 2005). According to the phenomenological perspective, I should like to stress how vital lived experience through the lived body is to build relationships, with plants and non-human animals, as well as other humans. This phenomenological aspect strongly links to ethical values of relationships and how we empathize with *the other*, whether the other is human or non-human. In this chapter, particular attention is given to foregrounding plants as historical and social individuals given their long lifespans on earth, and given their successful strategies for survival (Manetas 2010).

Being-in-the-world is obviously a subject–environment interaction. The human social system has always used ecological systems to survive in different ways. Different societies influence the population's attitude and behaviours towards nature (including plants), thus the environment is modified for the population's purposes. At the same time we affect the environment in many ways. The phenomenological position sees the physical subject-environment issue from an ethical point of departure. This may be interpreted as nature being a subject with its own agency, with its own life-worlds, experiences and relations. As such, plants should be treated with respect and the environment as an equal intersubjective partner.

Horizon

We are placed in the world as agents of opportunity (Alvesson and Sköldberg 2008). The opportunity lies in the freedom to choose, but at the same time we have an obligation to do so. But our options are limited. Therefore we have to consider what decisions might be appropriate to change from one condition to a new one. Decisions are made among other people, in a specific context, as a consequence of being social and cultural individuals. Society is always more than single, separate individuals, and it is more than a collective. A discussion concerning humanity's limitation of the non-human world is required owing to the ethical values of sustainable relations.

Every experience has its horizons directed at the past, the present and the future, which Merleau-Ponty (1995) refers to as "lived time". Lived experience involves lived time, which is experienced through the lived body, thus the horizon is part of the lived body. People experience situations differently on the basis of their personal horizons of experience. When we share experiences we are given the opportunity to move and expand our horizons. However, we need to acquaint ourselves with the other person's horizons. To do so, we proceed from preconceptions, rooted in our core values, in our life-world. When we understand one another we link the other person's experience, or horizon, with our own, and a fusion of horizons results (Gadamer 1997). Preconceptions form the basic conditions for comprehension, but they are also an obstacle to understanding and the origin of misunderstandings. Fusions of horizons lead to a revision of existing preconceptions. To deeply understand non-humans we have to share some of their lived experience and be a part of their lived time so as to expand and synthesize our horizons. Such shared experiences might assist a modification of human preconceptions and prejudices, in this case concerning plants.

When we strive for the horizon, we are able to discover our *capacity horizon* and thereby force the limits of what is conceivable (Berndtsson 2001). Discovering our capacity horizon makes us reach a *horizon of action*, which indicates a move towards increased knowledge and competence. Every person can thus continue discovering new capacity horizons, reaching new acting horizons and gaining increased understanding and competence. This is, according to Berndtsson, a precondition for lifelong learning. I would argue that it is also a precondition for an understanding of sustainability, and people's ability to change their ways of thinking and living in relation to the non-human world.

Ethical Phenomenology and Experience

As a vast philosophical tradition, phenomenology seeks to explain how we experience and perceive the world. This includes questions about individuals' relationships with the environment and how we interact with it. One of the most elementary kinds of experience we know is the one of value (Sanders and Wisnewski 2012). Such an experience plays a critical role in any effort to cope with ethical issues. It is in the eliminable feature of both the human condition and experience that we think through moral issues (ibid.). Thus phenomenology is important to ethical theorizing and problem solving.

As stated above, we are *in* the world; we are involved in it, concerned with it and, in fact, immersed in it (Heidegger 1993)—not as spectators to a spectacle but as participants. Therefore our relationships with the environments of the world inevitably involve ethical questions about what to do and how to do it (Sanders and Wisnewski 2012). Schutz (2002) suggests that in all situations we see, hear and act within value considerations. Our actions and attentions are never value free (Husserl 1989). According to this argumentation, people's and non-humans' experiences need to be taken seriously if we are to be sensibly involved with ethical matters.

One example of ethical experience is the encounter with "the other": the face-to-face interaction (Levinas 1979). Levinas claims that this meeting evokes a sense of responsibility towards the vulnerability of the other. You become aware of your obligations and you might choose not to fulfil them, but you know what you ought to do. It is an ethical decision to accept the other for their otherness, according to Levinas. Bridges can be built between you and the other through empathy, which means an open acceptance. Three foundational concepts of ethical phenomenology are otherness, responsibility and relationality (ibid.). Levinas also argues that otherness is essential to the world.

Thus far the phenomenological tradition of exploring experience has been human centred. What about the non-human world? And what about the ethics of encountering the non-human world and the intersubjective experiences of plants? Meeting with otherness should include the non-human world. Phenomenology elaborates on compound explanations of awareness. These will vary among different species and might depend on the consciousness of each. Examples of awareness that are significant to non-humans could be spatial awareness, attention, self-awareness, embodied action, intention in action, awareness of others and social interaction (Sanders and Wisnewski 2012).

Nature as Otherness and Plants as Objects

The ethical view posits a holistic perspective, yet we appear not to include plants in this holistic way of looking at the world. Plants have marked the evolution of life on earth and remain to play a crucial role in a balanced ecology (Tompkins and Bird 1989). Most life on earth depends on plants. Being the first link in most chains of life, plants have shaped environments and habitats for mammals and other animals for millions of years. We would simply not survive without plants.

In various ways, plants have played a significant role as objects in the story of human development. For example, they have been used for medicinal purposes, as a source of nutrition, and for cultural practices and rituals. Plants have also served as the basic components of various commodities, such as building materials, clothing, fossil-based energy sources and renewable energy (e.g. biofuels; Simpson and Ogorzaly 1995; Halivand et al. 2006; Imhoff et al. 2010).

In addition, plants play a broad and multifaceted role in our well-being: aesthetically, mentally and psychologically. An increasing interest in gardening and floristry provides evidence for this claim (Nyholm 2009; Park and Mattson 2009). However, and paradoxically, despite our obvious dependence on plants, we seem to have developed so-called "plant blindness", a concept which refers to our (1) incapacity to see or observe plants in our environment; (2) failure to understand plants' function on earth; (3) lack of appreciation of their aesthetic value; and (4) anthropocentric ranking of plants as inferior to animals (Wandersee and Schussler 1999).

One explanation for this phenomenon stresses human physical limitations, which indicates that we perceive and process only a fraction of what we see in the environment, and that we, perhaps unintentionally, pay attention to (1) movement; (2) bold colours; and (3) potential danger (Wandersee and Schussler 1999; Allen 2003; Manetas 2010). Yet, with this declaration based on our physical limitations, we are given an excuse to continue ignoring plants, thus it is normal and unavoidable for humans. These biological arguments are insufficient for several philosophical reasons, as noted by Merchant (1982), Plumwood (1993 and Hall (2011), as discussed later in this chapter. Another cause of plant blindness is a so-called zoocentrism, especially in a science-teaching context (Wandersee and Schussler 1999; Uno 2009; Manetas 2010; Balas and Momsen 2014). This means that teachers give animals priority over plants, and that, at all levels, they use animals to explain basic biological concepts (Uno 2009). Plants thus appear to be marginalized in relation to humans and other animals (Hallé 2002) and are seen simply as the backdrop to animal life.

Yet another explanation, albeit subtle, is the influence of patterns of thinking (Plumwood 1993; Hall 2011). Hall argues that zoocentrism is considered to be a method of excluding plants from human moral issues. Plumwood (1993) contends that the hierarchical order that places the plant kingdom lower than animals and humans is inherited from traditional Western ancient philosophy. Both Hall and Plumwood point out Plato's dualistic thinking and its influence on contemporary thought, and how this can give us further understanding of how plant blindness may have originated.

The otherness of nature is built on the notion of nature as an antagonist of human beings (Alexander 2013). Alexander's aim is to think beyond the anthropocentric point of view, and in so doing she raises two challenging ethical questions: Does nature belong to human beings? Or do human beings belong to nature? One related issue is how we construct "nature". Nature can include everything from the smallest particle to the whole macrocosm, or it can be defined by the term "all that grows". To the public in general, nature seems primarily to mean the physical, biological world, including biological diversity of the environments of forests and land (Sjöblom 2012). If cultural processes, including language, thinking patterns and practices, affect how we view the world, and especially how we value the non-human world, then we should look for alternative ways to organize knowledge of our environments (Balding and Williams 2016). "If immersed in a plant affiliated culture, the individual will experience language and practices that enhance capacity to detect, recall and value plants, something less likely to occur in zoo-centric societies" (p. 9).

As pointed out, our Western construction of nature derives to a great extent from the Ancient Greeks. Plato's dialectical world builds on a hierarchical order, which keeps nature in a subordinate position. This philosophy has led to a serious omission of plants, based on human interpretative prerogatives and domination of nature (Warren 2000). Merchant (1982) describes how ancient philosophy sees nature as the cause of disturbance and violence, and therefore believes that it must be tamed and held back. These thoughts have provided the basis for a longstanding idea—that of dominion over nature. A consequence of such a perception is the systematic exploitation and denial of plant consciousness (Hall 2011) which can be described as a process of bio-communications in plants, and that plants are sentient organisms. Plant blindness becomes normalized.

According to Levinas (1979), humans resist *the other* as *the other*. He suggests instead the *necessity* of otherness. He argues that we reduce *the other* into a part of what he calls *the same*. Looking at nature through the lens of Levinas' perspective, plants are valuable for their otherness thus their diversities are significant and meaningful because of the diversity itself. In addition, Hailwood (2000) criticizes the notion of a non-anthropocentric philosophy in which human and non-human distinctions are toned down. Instead, he stresses such distinction need to be valued for their otherness. But how can we create a shift in perspective to bring plants to the forefront and to challenge plant blindness?

Balding and Williams (2016) argue that both cultural and perceptual issues shape the ways in which we comprehend and value plants. They suggest that conservation programmes may promote identification and empathy with plants. The reason behind their proposal is both that conservation initiatives are biased against plants and towards animals, and that plants compose the majority of endangered species in Australia and other countries. In addition, Balding and Williams conclude that the constant harm and loss of plant diversity is ignored by politicians, and that humans' sense of connection with plants plays a crucial role in funding and support for the conservation of plants.

Place and Plants as Objects

The phenomenological perspective *object* is recognized as a phenomenon that is experienced by someone. Objects provide us with knowledge of the world. By using objects we are capable of understanding them, but our main interest is not to understand the object itself but what we can achieve by using it (Heidegger 1993). We use objects in order to explore the world and as tools to assist us.

An object that Fors (2003) describes from a life-world perspective is the room. People experience and relate to a room in various ways. Different types of room allow for different actions. Rooms mediate atmosphere and frame of mind owing to their design, furnishings, space, ongoing activities and so forth. Rooms are experienced both internally and psychologically, as well as an external physical object, as a *lived* room. A room can be a demanding enforced object, where we have been thrown in—,what Heidegger (1993) describes as "thrownness". To counteract the feeling of thrownness one must endeavour to inhabit the room, and thus be confident and secure in it. In contrast, pupils need to experience places in (nature) or: natural environments as their own lived room in order to relate to diverse natural environments and to connect with plants. Plants in natural environments may thus been seen as objects that offer the pupil significant knowledge about the place of a natural environment and hence about the world. The notions of being-in-the-world and place are, according to Heidegger, inseparably linked. Perceptions, feelings and memories are contained within a place. Therefore relationships between places and individuals are specific and subjective, multifaceted and dialectical; places are affected by people and vice versa (Thornton 2008). From a phenomenological point of view, the concept of place contains space, time and experience. It makes individual's feel at home in the world and discloses how they define themselves (ibid). If we see a room as a place and natural environments as different rooms, we understand the significance of embracing a permissive attitude given that pupils require lived experiences in natural environments. Art-based environmental education is aimed at such experiences.

Art-Based Environmental Education

Art can be seen as a teaching and learning tool, and thus an object. As such, art may help us to understand other phenomena, such as ecological processes. Using art as a tool in school may offer a contribution to pupils' knowledge of the world, and we might achieve a certain goal through aesthetic work and processes. Art-based environmental education may thereby connect pupils' life-worlds with natural environments.

The foundation of art-based environmental education (AEE) is that artistic activities support sensitivity to the environment (Mantere 1995). AEE addresses environmental understanding and accountability by developing an open attitude towards sense perceptions by using artistic methods (ibid.). Mantere claims that art makes one see and understand through emotions and all senses, and that one has to take the time to recognize the unknown and unexpected. Eco-artist Fielder (http://www.ericafielder-ecoartist.com/) points out that science and technology have already given us the knowledge and tools to prevent further environmental destruction. In addition we need compassion and a sense of relationship with nature that motivate us to embrace environmental concerns. From a phenomenological viewpoint, experiences and relationships are crucial to meaning-making processes (Selvi 2009). In an educational context, Dewey (2009) emphasizes embodied and sensory experiences in all education.

The term "art-based environmental education" was coined in the 1990s by Mantere (1992). She claims that art may play a crucial role in environmental education based on five assumptions driven by an aesthetic understanding of life and the environment:

- **Learning based on experience.** Emotional expressions are highly valued and form a foundation for environmental awareness. Creative activity gives weight to mental images, reflection and self-evaluation.
- **Future-oriented focus**. In our aim for a better environment and lifestyle, aesthetics plays a decisive role in creating a positive vision of the future. Aesthetic design includes social, scientific, psychological, aesthetic and ethical aspects.

- **A part of public space.** Pupils need to be given room to express themselves, and their experiences, dreams and visions, in public through a variety of aesthetic expressions. They need to know that they have an influence.
- **A positive vision of a change in lifestyle.** The emphasis is on the quality of change, not the quantity. Aesthetic environmental work contains creativity, inventiveness, ingenuity and dignity.
- **Cross-border projects.** With knowledge from different fields and disciplines, several things can take place. "In this way it is also possible to develop the mythical, metaphorical and deep-level psychological levels of man's relationship with nature into a constructive resource, in which factual information achieves deeper meanings" (Mantere 1992, p. 2).

Central phenomenological concepts are clearly connected with the aforementioned items, notably self-experience, lived time, lived room, the horizon of possibility and the capacity horizon. According to Merleau-Ponty (1995), our experiences are made conscious as we live them. Mantere (1992) claims that experience based on art also includes sensory involvement, ethical issues and respect for nature. Thus AEE can be regarded as a phenomenological approach.

Van Boeckel (2013) discusses how AEE can facilitate pupils' meeting with nature. He argues that art may open their minds to nature and give them space to vent any fears—for example, about global warming and other ecological ongoing crises. Van Boeckel highlights two aspects of art in environmental educational contexts. First, we associate art and aesthetic processes with pleasurable learning. Second, art can support meaning-making and an understanding of the ecological disasters that are approaching us. Art can, from this perspective, be seen as a way of living and a way of understanding one's existence. Similar explanations of didactic contexts are found in the literature, such as in the work of Aulin-Gråhamn et al. (2004), Austring and Sørensen (2006), Bamford (2006), Aure et al. (2009) and Häggström (2015). Common among these authors is an understanding of art as meaning-making, and as a teaching and learning tool, not only as a school subject. The principal aim of bringing aesthetics and art into schools is not to provide some balance and beauty in a shaky and frightening world because that, per se, does not change the world. The reason is rather to encourage pupils to open up and be more sensitive and receptive, thus giving them the opportunity to develop the ability to endure, despite a turbulent and troubled world. Aesthetics is not just about beauty; it can be frightening, repellent and unsightly because it relates to emotions and all of the senses (Wickman 2006). Wickman examines the relationship between aesthetics, science and learning. He focuses on educational practices and the need for aesthetic experiences while learning science. He argues that aesthetics in science is not about making science more pleasant for pupils to embrace; rather, it is equally important to consider the undesirable side of aesthetic experiences. Of particular significance when we look at learning processes in relation to science is the importance of learning more about the role of the unpleasant side of aesthetics in learning. Furthermore, Wickman stresses that the connections between aesthetics and science are more than the differences between the two. These assumptions correspondingly relate to the phenomenological position and conceptions of life-worlds, embodiment and emplacement, as discussed earlier.

Art as a Teaching and Learning Tool

Numerous studies have been conducted on the impact of aesthetic processes on learning (Aulin-Gråhamn et al. 2004; Austring and Sørensen 2006; Bamford 2006; Aure et al. 2009). Meaningfulness, participation and self-confidence are recurrent keywords when describing the effects of art as a teaching and learning tool (Drotner 1991; Danielsson et al. 2010; Häggström 2013). As a teaching and learning tool, art is seen as scaffolding; ways of supporting pupils in their learning processes and promoting a deep level of learning (Bruner 1996). "Scaffolding" is here defined in a broad perspective: supporting knowledge of self, others and the world at large.

"Cultural democracy", a concept from the community arts movement in the 1960s, is experiencing a renaissance today as a result of our modern multicultural society. The ideas behind this movement go well together with UNESCO's four systems concerning sustainability: the biophysical, the economic, the social and the political. Cultural democracy encompasses a number of interrelated concepts. One has to do with consideration and respect for all life on the planet and how to create harmonious interrelations. A second is about encouraging actions based on fruitful visions (Adams and Goldbard 1995). This movement includes an understanding of "empowerment", which is described as the ability to act through active participation in creative processes (Pringle 2008). Empowerment is a driving force in AEE too. Embodiment and empowerment should go hand in hand: one cannot be empowered unless the experience is embodied, and only then will the experience become part of the life-world.

Land art is a creative and artistic approach to experience and it explores nature and natural materials. Making land art in schools is a way to learn about nature and natural places that combines creativity and learning (Pouyet 2009; Shilling 2009; Kastner 2012), and it offers the opportunity to challenge plant blindness. The demand for research on land art, environmental art and eco-art is growing, with increasing research in this field in Finland, the UK, Germany, the USA, Canada and Australia (Sørenstuen 2013). This has resulted in new university courses that are topical and future oriented. "Environmental art" is an umbrella term for numerous concepts such as land art, earth art, eco-art and restoration art. Connection with nature and the use of natural materials bring different types of environmental art together. It is common to relate environmental art to a specific place, and place-based research highlights the importance of places for both aesthetics and sense-making (ibid.). Land art can help pupils to see nature *as something*, and natural materials may support pupils' understanding of ecological processes, their place in the world and sustainable issues. Creating land art and involvement with natural materials makes room for the non-human world to some extent. This is an experience of the lived body, according to Merleau-Ponty (1995), thus I am considering the ways in which being physically close to natural materials in the natural environment might help pupils to feel a closeness to plants and to gain an embodied knowledge of place. The only way of understanding the world, Merleau-Ponty claims, is to experience it—namely, the lived experience. Plants thus become part of a pupil's lived sensory experience and part of their life-world. Land art opens up horizons (as discussed earlier) for ethical debate, and questions about materials and the environment. A crucial point here is the appreciation of the otherness of plants (Levinas 1979; Fig. 1).

Fig. 1 Examples of land art that everyone can make. Land art is about being outdoors, experiencing nature and exploring natural materials in natural places while having an aesthetic experience

Education and Sustainability

Education supporting a sustainable future was the main subject discussed at UNESCO's 2002 World Summit on Sustainable Development in Johannesburg. Sustainability is, as mentioned above, justified in terms of four interdependent systems: the biophysical, the economic, the social and the political. Connecting these systems to environmental concerns is a fundamental feature of sustainability. Building such connections, Fien (2004) claims, stresses a deep and determined way of thinking about education, which involves critical thinking and creativity. Creativity is here seen as problem solving and a way of rethinking in a holistic phenomenological way. This means that education will encourage a structure of "ethics and values that is sensitive to cultural identity, multicultural dialogue, democratic decision-making and the appropriate use and management of natural resources" (p. 8).

By order of Skolverket, a research project with the purpose of evaluating environmental education in Swedish schools was carried out in 2001 (Öhman 2004). Three different pedagogical approaches were identified as fact-based environmental education, normative environmental education and pluralistic environmental education (ibid.). The fact-based approach places great trust in scientific expertise as the solution of environmental problems. Schools' mission, then, is to disseminate scientific knowledge to students so that they can act accordingly. Central to the normative approach is the conflict between humans and nature, which leads to various environmental problems. Scientific knowledge about nature is seen as the solution, which will guide pupils in their decision-making regarding environmentally moral issues. In the last of the three pedagogical approaches, environmental problems are seen as conflicts between people, and problems are considered to be social constructions. Environmental discussion is connected with the development of society in general, and a pluralistic perspective is applied, such as the examples of various scientific, and sometimes conflicting, approaches (ibid). Wals (2015) is concerned by how education pay attention to global sustainability challenges. He stresses the need for education to involve emancipatory perspective and a view on sustainability competence which refers to competence as "relational, contextual and emergent property" (ibid p. 11). This involves knowledge, doing and being and how to tranform this into action. Education on sustainability can not be restricted to classrooms. It requires a mixture of informal, non-formal and formal education.

To increase and ensure quality environmental education, Skolverket (2002) recommends clear political signals, research on education for sustainable development, support for teachers and teacher educators, and support for schools in their efforts towards sustainable development. On the other hand, UNESCO (2014) suggests that we need a change in our way of thinking; it is not enough to develop new technical innovations and new political guidelines and legislation. According to UNESCO, education for sustainable development would benefit from embracing serious and difficult issues, such as climate change, biodiversity, catastrophe risk reduction, poverty reduction and sustainable consumption. This requires participatory pedagogy in order to encourage and empower pupils to take action. Critical thinking is therefore a key concept, and so is the ability to imagine future scenarios. Interaction and collaboration are seen as necessary. In relation to Öhman's three pedagogical approaches, Skolverket's recommendations can be seen as fact-based and normative environmental education, while UNESCO (2009; 2014) advocates a pluralistic environmental education. This difference may explain why environmental education varies in Swedish schools.

Concluding Thoughts

Plant blindness might be prevented when teachers and pupils engage with plants and natural environments. It is vital that children and young people have the opportunity to develop relationships with plants (Magntorn 2007; Sjöblom 2012; Nyberg and Sanders 2014). Nyberg (2008) points out that pupils' interest decreases over time, which is why continuity and repeated visits to natural environments may be essential to create a lasting interest. Accordingly, pupils need to be given time and room to stay and play in natural environments, learning *in* nature, not only *about* nature (Escamilla 2013; Golden 2013). To develop an understanding of science concepts and processes, lessons ought to be taken not only in science class but rooms elsewhere. From a phenomenological approach, I suggest that education concerning sustainability should pay more attention to the lived body and thus involve various aesthetic teaching and learning tools in school. The methods and theories of AEE are grounded in ethical aspects and environmental awareness, and they aim to empower pupils. From a phenomenological perspective, empowerment depends on embodiment—when embodied actions become a strong part of one's life-world. Land art, too, as an example of a combination of aesthetic and sustainable teaching and learning methods, can open up opportunities for deep ethical discussions of otherness, the non-human world and how to interact in a respectful way. Alexander's (2013) critical questions regarding whether humans belong to nature can be reformulated: are we not nature?

References

Adams, D., & Goldbard, A. (1995). *Cultural democracy: Introduction to an idea.* Seattle: Webster's World of Cultural Democracy.

Alexander, V. (2013). *Environmental otherness: Nature on human terms in the garden.* http://www.otherness.dk/fileadmin/www.othernessandthearts.org/Publications/Journal_Otherness/Otherness__Essays_and_Studies_4.1/Environmental_Otherness.pdf

Allen, W. (2003). *Plant blindness. BioSceince, 53*(10). https://www.deepdyve.com/lp/oxford-university-press/plant-blindness-d089roV6SF

Alvesson, M., & Sköldberg, K. (2008). *Tolkning och reflektion. Vetenskapsfilosofi och kvalitativ metod.* Lund: Studentlitteratur.

Aulin-Gråhamn, L., Persson, M., & Thavenius, J. (2004). *Skolan och den radikala estetiken.* Lund: Studentlitteratur.

Aure, V., Illeris, H., & Örtengren, H. (2009). *Konsten som läranderesurs. Syn på lärande, pedagogiska strategier och social inklusion på nordiska museer.* Skärhamn: Nordiska akvarellmuseet.

Austring, B., & Sørensen, M. (2006). Æstetik og læring. *Grundbog om æstetiske læreproces*ser. Köpenhamn: Socialpædagogisk bibliotek Hans Reitzels forlag.

Balas, B., & Momsen, J. L. (2014). Attention "Blinks" differently for plants and animals. *CBE—Life Sciences Education, 13*, 437–443.

Balding, M., & Williams, K. (2016). Plant blindness and the implications for plant conservation. *Conservation Biology, 30*(6). doi:10.1111/cobi.12738.

Bamford, A. (2006). *The wow factor. Global research compendium on the impact of the arts in education.* Münster: Waxman.

Bengtsson, J. (1988). *Sammanflätningar. Husserls och Merleau-Pontys fenomenologi.* Goteborg.Daidalos

Berndtsson, I. (2001). *Förskjutna horisonter. Livsförändring och lärande i samband med synnedsättning eller blindhet.* Doctoral thesis, Acta Universitatis Gothoburgensis, Göteborg.

Bruner, J. (1996). *The culture of education.* Cambridge, MA: Harvard University Press.

Claesson, S. (2009). *Lärares hållning.* Lund: Studentlitteratur.

Danielsson, H., Graviz, A., & Odelfors, B. (2010). *Multimodal representations and the construction of knowledge.* Paper to the conference for learning, Stockholms Universitet, Stockholm.

Dewey, J. (1958). *Experience and nature.* New York: Dover Publications.

Dewey, J. (2009). *Democracy and education. An introduction to the philosophy of education.* Lexington/Kentucky: Feather Trail Press.

Drotner, K. (1991). *At skabe sig – selv. Ungdom, æstetik, pedagogik.* Köpenhamn: Gyldendal.

Escamilla, I. M. (2013). Drawing, photographs and painting. Learning about the natural world in an urban preschool. In D. R. Meier & S. Sisk-Hilton (Eds.), *Nature education with young children*. New York: Routledge.

Fien. (2004). *Education for sustainability. From Rio to Johannesburg: Lessons learnt from a decade of commitment*. Paris: United Nations Educational, Scientific and Cultural Organization.

Fors, V. (2003). Science centrets unika förutsättningar – om att sätta lärandet i sitt sammanhang. I *Några nedslag i livsvärlden. Teori och praktik i pedagogisk forskning*. Luleå: Luleå tekniska universitet.

Gadamer, H.-G. (1997). *Truth and method*. New York: Continuum.

Golden, A. (2013). Preeschool children explore the forest. The power of wild spaces in childhood. In I. D. R. Meier & S. Sisk-Hilton (Eds.), *Nature education with young children*. New York: Routledge.

Griffiths, J. (1999). *PiP PiP*. London: Flamingo.

Häggström, M. (2013). Delaktighet och kommunikation genom visuell representation. In I. T. Barow (Ed.), *Mångfald och differentiering. Inkludering i praktisk tillämpning*. Lund: Studentlitteratur.

Häggström, M. (2015). Become your opposite person. Participation and communication through visual representation. In I. Â. Saldanha, C. Trigo, M. Pardiñas, & T. Torres de Eça (Eds.), *Risk and opportunity for visual art education in Europe*. APECV. isbn:978–989–99073-2-4.

Hailwood, S. (2000). The value of nature's otherness. *Environmental Values, 9*, 353–372.

Halivand, W. A., Prins, H. E. L., Walrath, D., & McBride, B. (2006). *The essence of anthropology*. Belmont: Wadsworth Publishing.

Hall, M. (2011). *Plants as persons. A philosophical botany*. New York: State University of New York.

Hallé, F. (2002). *In praise of plants*. Portland: Timber Press.

Heidegger, M. (1993). *Varat och tiden*. Göteborg: Daidalos.

Husserl, E. (1989). *Fenomenologins idé*. Göteborg: Daidalos.

Imhoff, M., Bounanza, L., Wolfe, R. E., & Zhang, P. (2010). Remote sensing of the urban heat island effect across biomes in the continental USA. *Remote Sensing of Environment, 114*(3), 504–513.

Kastner, J. (2012). *Nature*. London: Whitechapel Art Gallery.

Levinas, E. (1979). *Totality and infinity*. New York: Springer.

Magntorn, O (2007). *Reading nature. developing ecological literacy through teaching.*. Doctoral thesis, Linköpings universitet, Linköping.

Manetas, Y. (2010). *Alice in the land of Plants. Biology of plants and their importance of planet earth*. Heidelberg/New York: Springer.

Mantere, M. H. (1992). Ecology, environmental education and art teaching. In L. Piironen (Ed.), *Power of images* (pp. 17–26). Helsinki: INSEA Research Congress, Association of Art Teachers.

Mantere, M. H. (1995). *Image of the earth: writings on art-based environmental education* (pp. 18–28). Helsinki: University of Art and Design.
Merchant, C. (1982). *The death of nature*. London: Wildwood House.
Merleau-Ponty, M. (1995). *Kroppens fenomenologi*. Göteborg: Daidalos.
Nielsen, C. (2005). *Mellan fakticitet och projekt. Läs- och skrivsvårigheter och strävan att övervinna dem*. Doctoral thesis, Acta Universitatis Gothoburgensis, Göteborg.
Nyberg, E. (2008). *Om livets kontinuitet. Undervisning och lärande om växters och djurs livscykler – en fallstudie i årskurs 5*. Doctoral thesis. Acta Universitatis Gothoburgensis, Göteborg.
Nyberg, E., & Sanders, D. (2014). Drawing attention to the 'green side of life'. *Journal of Biological Education, 48*(3), 142–153.
Nyholm, C. B. (2009). *Potted flowers and plants may help speed hospital recovery*. Retrieved from: http://www.associatedcontent.com/article/1352323/potted_flowers_and_plants_may_help.html
Öhman, J. (2004). Moral perspectives in selective traditions of environmental education – Conditions for environmental moral meaning-making and students' constitutions as democratic citizens. In P. Wickenberg et al. (Eds.), *Learning to change our world?* Lund: Studentlitteratur.
Park, S. H., & Mattson, R. H. (2009). Ornamental indoor plants in hospital rooms enhanced health outcomes of patients recovering from surgery. *The Journal of Alternative and Complementary Medicine, 15*(9), 975–980.
Plumwood, V. (1993). *Feminism and the mastery of nature*. London: Routledge.
Pouyet, M. (2009). *Natural. Simple land art through the seasons*. London: Frances Lincoln Publishers.
Pringle, E. (2008). Kultur, Lärande och konstpedagogikens föränderliga roll. In*Konst och Pedagogik*. Gävle: Gävle konstcentrum.
Sanders, M., & Wisnewski, J. (Eds.). (2012). *Ethics and phenomenology*. Plymouth: Lexington Books.
Schütz, A. (2002). *Den sociala världens fenomenologi*. Göteborg: Daidalos.
Segolsson, M. (2011). *Lärandets hermeneutik. Tolkningens och dialogens betydelse för lärandet med bildningstanken som utgångspunkt*. Doktorsavhandling, Högskolan i Jönlöping, Jönköping.
Selvi, K. (2009). Phenomenological approach in education. In A.-T. Tymieniecka (Ed.), *Education in human creative existential planning* (Vol. 95, pp. 39–51). Dordrecht: Springer.
Shilling, R. (2009). *Land art*. http://richardshilling.co.uk/section348741.html
Simpson, B. B., & Ogorzaly, M. C. (1995). *Economic botany: Plants in our world*. New York: McGraw-Hill.
Sjöblom, P. (2012). *Naturen och jag. En studie av gymnasiestuderandes förhållande till naturen ur ett miljöpedagogiskt perspektiv*. Doctoral thesis, ÅboAkademi University Press, Åbo.

Skolverket [Swedish National Agency for Education]. (2002). *Skola för hållbar utveckling. Miljöundervisning och utbildning för hållbar utveckling i svensk skola*. Stockholm: Skolverket.

Sørenstuen, J.-E. (2013). *Levande spår – att upptäcka naturen genom konst och konsten genom natur*. Lund: Studentlitteratur.

Thornton, T. F. (2008). *Being and place among the Tlingit. Culture, place and nature*. Seattle/London: University of Washington Press.

Tompkins, P., & Bird, C. (1989). *The secret life of plants. A fascinating account of the physical, emotional, and spiritual relations between plants and man*. New York: Harper and Row.

UNESCO [United Nations Educational, Scientific and Cultural Organization]. (2009). *Review of contexts and structures for education for sustainable development*. Paris: United Nations Educational, Scientific and Cultural Organization.

UNESCO [United Nations Educational, Scientific and Cultural Organization]. (2014). *Shaping the future we want. UN decade of education for sustainable development (2005–2014) FINAL REPORT*. Paris: United Nations Educational, Scientific and Cultural Organization.

Uno, G. E. (2009). Botanical literacy: What and how should student learn about plants? *American Journal of Botany, 96*(10), 1753–1759.

van Boeckel, J (2013). *At the heart of art and earth. An exploration of practices of art based environmental education*. Doctoral dissertation, Aalto University Publication Series, Aalto.

Wals, A. (2015). *Beyond unreasonable doubt. Education and learning for socio-ecological sustainability in the Anthropocene*. Wageningen: Wageningen University.

Wandersee, J. H., & Schussler, E. E. (1999). Preventing plant blindness. *The American Biology Teacher, 61*, 84–86.

Warren, K. (2000). *Ecofeminist philosophy: A western perspective on what it is and why it matters*. Lanham: Rowman and Littlefield Publishers, Inc.

Wickman, P.-O. (2006). *Aesthetic experience in science education: Learning and meaning- making as situated talk and action*. Doctoral thesis, Lawrence Erlbaum Associates, Mahwah.

Challenging and Expanding the Notion of Sustainability Within Early Childhood Education: Perspectives from Post-humanism and/or New Materialism

Kassahun Weldemariam

Introduction

Drawing on post-humanism and/or new materialism as a theoretical framework, this chapter challenges the notion of sustainability within education in general (the way it is understood, described and discussed) and within early childhood education for sustainability (ECEfS) in particular (Davis and Elliott 2014). To offer an understanding of sustainability within existing bodies of knowledge, the chapter focuses on various aspects of the sustainability journey in early childhood education (ECE), including historical beginnings, policy discourses, curricular design, pedagogical practices and research approaches within ECEfS.

The chapter begins with an overview of the historical trends and associated macrolevel/international dialogues pertaining to sustainability. This is followed by a section that describes curricular representation and pedagogical practices, and reviews prior research on sustainability within the field of ECE as well as the underpinning ideas. The subsequent section provides an introduction to post-humanism/new materialism as a theoretical

K. Weldemariam (✉)
Department of Pedagogical, Curricular and Professional Studies,
University of Gothenburg, Gothenburg, Sweden

O. Franck, C. Osbeck (eds.), *Ethical Literacies and Education for Sustainable Development*, DOI 10.1007/978-3-319-49010-6_7

framework and how these theories provide stimulus for rethinking the ontological, epistemological and ethical conditions underpinning sustainability education. The next section critically discusses and challenges the prevailing notion of sustainability through the lens of post-humanism/ new materialism and attempts to elucidate its potential for addressing, challenging and expanding the notion of sustainability. I conclude with a proposed way forward for ECEfS.

Historical Beginnings in Sustainability Education and its Underpinning Policy Discourses

Despite the existence of diverse views about the origin of the term "sustainability", its genesis and history are often associated with nature conservation education, nature study and environmental education in the 1960s (Wals 2012; Somerville 2015). Nature conservation education broadly emphasizes educating citizens to understand, appreciate, connect with and protect nature. Building on nature conservation education, the overall purpose of environmental education has been to address the integration of environmental issues into formal education, with a view to influencing citizens' environmental behaviour and enabling them to live in conscious recognition of the earth's carrying capacity (Wals 2012).

The recognition of the link between environmental issues and ECE dates back to the 1990s. According to Tilbury (1994) and Palmer (1995), as cited in Davis and Elliott (2014), it is the recognition of the unique affordances of children's curiosity that has led to the identification of ECS as a foundation for lifelong learning and the development of pro-environmental values and attitudes (Davis and Elliott 2014). The movements associated with the United Nations Convention on the Rights of the Child (UNCRC) and UNESCO's (2005) Decade of Education for Sustainable Development (DESD) are other significant events that played an important role in promoting children as social agents with rights to participate in matters relevant to them—including environmental issues.

However, despite this recognition and movement, the notion of sustainability had not been overtly incorporated into ECE policy frameworks and pedagogical practices until relatively recently. UNESCO's first official report on the subject, *The Contribution of Early Childhood Education for a Sustainable Society* (Pramling Samuelson and Kaga 2008), was an initiative to explicitly address sustainability within ECE. This report contributed to an increased interest in the concept of sustainability within early childhood pedagogy, curricula and research. Building on the work of UNESCO, the World Organisation for Early Childhood Education (OMEP) made a plea for the expansion of the field by highlighting the link between ECE and sustainability (Siraj-Blatchford et al. 2010).

Likewise, by indicating the intergenerational nature of sustainability, the Brundtland Report (1987) has also played a significant role in, and has emphasized the need for, lifelong engagement and commitment, which implies the need to practise the skills of sustainability in terms of social justice and equity. This has led to the inclusion of sustainability and its accompanying dispositions (care, ethical responsibility and empathy) as a foundation for lifelong learning in early years education.

Curricular, Pedagogical and Research Approaches Within Early Childhood Education for Sustainability

Although there are commonalities across nations, different curricular frameworks are designed on the basis of different philosophies, theoretical underpinnings, beliefs and values. Most early childhood curricular frameworks have been designed based on philosophies such as child-centredness, socio-constructivist learning theories, activity-based learning, experiences and situations, and immersion and interaction with "nature" and the physical environment (Vygotsky 1986; Lave and Wenger 1991; Piaget 1997). Drawing on these philosophical underpinnings, various early childhood scholars across the world have explored different curricular and pedagogical approaches that are intended to transform ECE practices relating to sustainability.

One main approach has been the focus on in-service teachers, with the aim of enhancing their participation and ability to critically reflect on, and play a leading role in, children's engagement with sustainability issues within ECE settings (Ärlemalm-Hagsér 2014; Engdahl and Ärlemalm-Hagsér 2014; Ji and Stuhmcke 2014; Mackey 2014; Phillip 2014; Young and Cutter-Mackenzie 2014). Others (O'Gorman 2014; Sundberg and Ottander 2014; Gilbert et al. 2014) argue for the need to strengthen preservice teacher education by emphasizing critical reflection ability within the training programme.

Another popular pedagogical/curricular approach to sustainability is the project approach, which involves a deeper and topic-based exploration of sustainability as a mechanism to engage children with pertinent issues.

Through the project, children are encouraged to engage with different aspects of sustainability and solve local problems within their community, while learning about sustainable practices in the process (Engdahl and Ärlemalm-Hagsér 2014; Ji and Stuhmcke 2014).

Additionally, outdoor education, nature-based studies and place-based pedagogy are approaches that are promoted for the opportunities they provide in creating a comprehensive context for connecting children with nature (Miller 2014; Mackey 2014; Engdahl and Ärlemalm-Hagsér 2014; Barratt et al. 2014; Chawla and Rivkin 2014; Sundberg and Ottander 2014; Gilbert et al. 2014). It is believed that such a context lays the foundation for children's understanding of, engagement with and enactment of sustainability.

Researchers have also emphasized the importance of home/school partnership, children's community experience and active participation as citizens (Barratt et al. 2014; Chawla and Rivkin 2014; Engdahl and Ärlemalm-Hagsér 2014; Ji and Stuhmcke 2014; Mackey 2014; Phillip 2014; Young and Cutter-Mackenzie 2014). They indicate the need for those involved in ECE to facilitate and organize such opportunities for children.

Other recurring pedagogical approaches and related discourses include addressing sustainability as learning content and identifying specific behaviours; values or attitudes; environmental learning; education in, for and about the environment; advocacy and children's rights (Engdahl and Rabusicova 2011); teachers' pedagogical competence and implementation of education for sustainable development (ESD), with a shift from literacy to action (Hedefalk et al. 2014); and play-based pedagogy for environmentalizing early childhood curricula (Edwards and Cutter-Mackenzie 2011). Phillip (2014), on the other hand, has employed storytelling as a pedagogical tool for engaging children with sustainability.

Throughout the aforementioned curricular and pedagogical approaches, a recurring dominant discourse is the perception of the child as an active agent and critical problem-solver who is visible and able to actively engage and make decisions. The notion of agency has also been evident in international discourses, as indicated by the UN Millennium Summit, which described children as central actors and critical agents of change, who have infinite capacities for activism to create a better world (UN 2015).

Regarding theoretical orientations of previous studies within ECEfS, positivist, interpretivist, critical-theory and rights-based approaches have been identified as the most widely used research paradigms (Somerville and Williams 2015). Studies that employ a positivist approach (Kahriman-Öztürk et al. 2012; Hadzigeourgious et al. 2011) tend to quantify children's attitudes, beliefs, perception and achievement in environmental education and outdoor learning spaces through pre- and post-test knowledge.

In particular, scholars who employ the interpretative paradigm are mainly situated within the dominant theoretical discourses of "connection to nature" and "children's rights" (Gambino et al. 2009). Emanating from the legacy of Rousseau, studies that are situated within the "connection to nature" discourse are mainly related to "green" environmental issues and children's alienation from nature is their main concern; consequently, they advocate the need to reconnect children to nature (Taylor 2013).

On the other hand, scholars such as O'Gorman and Davis (2012) employ a critical perspective that is intended to bring about change through the participation and involvement of children in research and practical activities. They examined children's and teachers' responses to the use of an ecological footprint calculator as a sustainability pedagogical tool. Likewise, Ärlemalm-Hagsér (2013) employed critical theory and ecofeminist perspectives to examine preschool children's agency and meaning-making as integral to sustainability pedagogies. Her study indicated the potential of preschool as a transformative arena for the expression of different political and pragmatic agendas.

In order to briefly overview the current and prevailing research discourses within ECEfS, I have attempted to summarize the content of the available research reviews, international research collections and individual articles in the field from 1996 to 2015, to offer a fairly comprehensive and representative picture of the main discourses on the subject. In so doing, I present the key terms: childhood and sustainability; early childhood and environmental education; early childhood; and ESD. As part of my research I consulted articles, books and book chapters and policy reports within and beyond different databases: Scopus, ERIC, Web of Science and Google Scholar. Despite the effort to present a reasonably comprehensive coverage of the current and leading ideas on sustainability within early childhood studies, I acknowledge that there are "missing voices" due to the limited scope and nature of the chapter.

Table 1 summarizes key research reviews and international research collections, highlighting the themes they address. The individual articles (most of which are also embraced within the list of reviews in table 1) are used and cited throughout the chapter. To indicate how the discourse has evolved over time, the reviews are presented in chronological order, from the oldest to the most recent.

Table 1 Thematic summary of review articles and international research collections within ECEfS

Journal articles/book	*Themes addressed*
Barratt Hacking, et al. (2007). Engaging children: Research issues around participation and environmental learning. *Environmental Education Research, 13*(14), 529–544	Engagement of children as environmental stakeholders, environmental learning, outdoor activity and children's well-being
Davis, J. (2009). Revealing the research 'hole' of early childhood education for sustainability: A preliminary survey of the literature. *Environmental Education Research, 15*(2), 227–241	Children's relationship with nature (education in the environment) and the understanding of environmental topics (education about the environment) is the focus Lack of research work on children as agents of change (education for environment) Time frame addressed: 1996–2007
Davis, J., & Elliott, S. (2014). *Research in early childhood education for sustainability: International perspectives and provocation.* London: Routledge	Young children's actual and potential capabilities as agents of change for sustainability Participatory, socio-constructionist and systemic approaches towards social transformation Children as social agents, critical thinkers and problem-solvers who are able to act in collaboration with the community
Hedefalk, M. et al. (2014). Education for sustainable development in early childhood education: A review of the research literature. *Environmental Education Research (Journal Article), 21*(7), 1–16	Explicit focus on ESD ESD is described in two different ways: (1) education about, in and for the environment (2) as an approach to education including economic, social and environment pillars Teachers' understandings of ESD and how it can be implemented, curricular integration and pedagogical adaptation A shift from literacy to action-oriented education (i.e. children as competent actors) Time frame addressed: 1996–2013
Somerville, M., & Williams, C. (2015). Sustainability education in early childhood: An updated review of research in the field. *Contemporary Issues in Early Childhood, 16*(2), 102–117	Critical analysis and categorization of earlier research based on theoretical and methodological orientations, and identification of three main theoretical orientations: connection to nature, children's rights and the post-humanist framework

As can be seen from the table, central and recurring themes within ECEfS research include environmental awareness; understanding and active engagement with the environment; environmental and outdoor learning; participatory learning; relationship with nature; ESD as content-based learning; action-oriented practices; and children's agency. These areas tend to emphasize knowledge-building and behavioural change for a sustainable future, that is, they are human/child-centred and rely on the cognitive and meaning-making processes of the autonomous and learning child. Building on Somerville and William's (2015) recommendation on the potential of the post-humanist approach for addressing sustainability within ECEfS, I attempt to deconstruct and expand the aforementioned notion of sustainability from a post-humanist/new materialist perspective which embraces the agency of the non-human world.

Post-humanism and/or New Materialism as a Theoretical Framework

Post-humanism and new materialism are contemporary philosophical movements/theoretical frameworks that significantly overlap with one another, and there is no definitive distinction between them. Different scholars define and describe them in different ways. Some scholars (Braidotti 2013; Lenz-Taguchi 2014; Taylor and Hughes 2015) consider new materialism as a field that comes under the broader post-humanism line of thinking, while others (DeLanda 2008; Dolphijn and van der Tuin 2012) tend to treat new materialism as its own field of study. This chapter is not concerned with the debate about the similarities and differences between the two fields. Rather, it draws on their significant commonality and how they can help us to rethink sustainability in ECE.

In a broader sense, the post-humanism framework departs from humanism and accepts the humanistic premise of critiquing transcendent explanations of human existence, but it redefines the human as a part of (not separate from) the natural world, such that human nature is a multispecies and entangled event, that is we humans live in a "common world" with others and we are made up of our intra-actions with nonhumans (Latour 2004; Barad 2007; Mickey 2007, 2016; Haraway 2008; Braidotti 2013; Taylor 2013). New materialism accepts the materialistic premise of critiquing idealistic explanations of self and world and replacing them with materialistic explanations, but it redefines matter in process-relational and active/agential terms in contrast to classical and modern views of matter as substantial and passive (Bennet, 2010; DeLanda 2008; Lenz-Taguchi 2010; Mickey 2010; Dolphijn and van der Tuin 2012).

Post-humanism and new materialism challenge the long-standing idea of looking at the human subject as an exceptional and political agent, and states that such privileging of the individual human subject is problematic, especially when confronted with those political problems that seem tied to, and around, the subject's very identity and anthropocentric actions (Poe 2011; Braidotti 2013). Both approaches promote the notion of humanity as embedded within a more-than-human network context—not alone as a sole agentic force in the world—and argue that disregarding the agency of the more-than-human in today's Anthropocene era means that humanity remains stuck in its own myth of exceptionalism, at the political cost of the continued human dominance over the environment and other entities (Delanda 2008; Poe 2011; McKenzie and Bieler 2016). In this way, both frameworks problematize anthropocentric thinking, which considers humans as the central concern and the sole bearer of agency, and instead recognize distributed agency among humans and more-than-human others (Deleuze and Guattari 1987; Latour 2004, 2005). The term "Anthropocene" denotes the current geological age during which human activity has been the dominant influence on the climate and the environment. Using the idea of the Anthropocene, these frameworks reconsider the relationship between the human and the physical/material world by re-imagining a new paradigm, which repositions humans from the perspective of their embeddedness in interdependent socio-ecological systems (Malone et al. (2017); Somerville 2015).

The two frameworks present different modes of being and knowing, wherein both human and more-than-human aspects of the world are positioned in a "flat" ontology without any centre and hierarchy (Deleuze and Guattari 1987). Their epistemology is concerned with non-human experience as a site of knowledge (Taylor and Hughes 2015) and with ethical conditions that take into account a broader and interconnected notion of life, embracing the non-human/more-than-human world and other forms of life (Wolfe 2010). According to Taylor and Hughes (2015), "thinking posthuman ethics begins by re-thinking interdependence, by including nonhumans in an ethics of care, by understanding the human always and only in-relation-to nonhumans who are no longer 'others' but are, intimately and always, ourselves as the body multiple" (p.15). Parallel to this, Taylor (2013) raises intergenerational and interspecies justice, inheritance and responsibility as fundamental ethical considerations of sustainability in the contemporary anthropocentric era.

Early Childhood Education for Sustainability Through the Lens of Post-humanism/New Materialism

As briefly described at the beginning of this chapter, the historical development of ECEfS aimed at building a foundational knowledge with children so that they demonstrate care, ethics and empathy towards "nature" and the environment. From a post-humanism/new materialism perspective, such an approach tends to be human-centric as it focuses on the need to cultivate human knowledge, skills and attitudes towards environmental stewardship, that is, human-oriented education for environmental sustainability (Taylor 2013, 2017).

This human-centric aspect also featured in the Brundtland Report (1987), which is a landmark policy document that has led to the inclusion of sustainability skills such as lifelong engagement and commitment in early years education. In the report, sustainable development is defined as "development that meets the needs of the present without compromising the ability of future generations to meet their own needs" (1987, p. 16). Even though the document provides a basis for today's developed and complex understandings of sustainability and sustainable development, this definition has anthropocentric features as it is mainly concerned with enabling one to fulfil the needs of the present "human" generation without compromising the possibilities of future "human" generations. The definition tends to imply separation between humans and the physical world by putting humans at the centre and emphasizing the finite resources of the environment, with an intergenerational focus and intergenerational equity aspect. Owing to its anthropocentric nature, the notion of sustainable development, as described in the Brundtland Report, tends to focus more on how to develop and sustain the fulfilment of human needs rather than how to sustain the biosphere per se, which is a prerequisite for sustainable development.

As a lead organization in relation to matters such as sustainable development, UNESCO has been the driving force in shaping policies pertaining to sustainability education in general and ECEfS in particular. However, sustainability, as articulated by UNESCO, tends to be limited by human-centric views that emphasize not only the ways in which humans damage the environment, but the ways in which humans need to become environmental stewards. As pointed out by Taylor (2013, 2017), there is a need to perceive the human–environment relationship beyond stewardship. The notion of outdoor education as stated in Section IV of Agenda 21 of the UN (1992) is an instance of UNESCO's human-centred discourse. This document perceives and depicts humans as either villains or heroes and considers agency as an exclusively human attribute. This view tends to neglect the agency of the more-than-human world. Challenging such a notion from a post-humanism/new materialism perspective raises "an important issue" within the environmental education and sustainability discourse, that is, it heavily relies on human exceptionalism while we humans have much to learn from and with other species and the non-human world.

Hence, from a post-humanist/new materialist perspective, the human-centric way of ontological positioning tends to be a reductionist approach as it promotes the nature/environment/earth as something to be controlled, saved or mastered, by human actors. Post-humanist approaches to environmental relations tend to emphasize the mutually constitutive and entangled relationships between humans (as just one species among many) within a common world (Latour 2004; Taylor 2013). Since this approach problematizes anthropocentric thinking, it can serve as an important theoretical tool to allow us to rethink human relationships with the environment and the physical world. This way of understanding focuses on agency as something that comes out of relationships and assemblages (including human and more-than-human) and not just as a human attribute (Latour 2004, 2005).

Thus thinking in terms of post-humanism/new materialism within ECEfS helps to reveal the limits of the "dominant" conceptions of the human–environment relationship within the history and policy documents pertaining to ECEfS. I argue that these conceptions have widened the gap between humans and the physical world, and that post-humanism/new materialism allows a rethinking of our being and humans' relationship with the environment. The historical foundations and UNESCO policy documents mentioned earlier have a political power and shape our understanding of the notion of sustainability in early childhood and beyond, and hence it is worth examining the way in which they depict the human–environment relationship, which in turn helps us to revisit our ontologies, epistemologies and educational practices.

Curricular and pedagogical approaches within ECEfS were addressed earlier in this chapter. As highlighted previously, the recurring curricular and pedagogical themes within ECEfS include the outdoors as a learning environment; teacher training and competence; place-based pedagogy; the project approach; storytelling; children's rights and community participation; advocacy; relationships and interactions with nature; the recognition of the child's uniqueness; the notion of agency; and the value of partnerships between home and the early years settings. The post-humanism/new materialism perspective challenges such approaches and extends the discourse beyond children's agency, to explore what can be learnt by repositioning humans in a flat ontology where agency is produced as a constituent element of human and non-human entities.

As pointed out by McKenzie and Bieler (2016, p. XIII), ever since the term "sustainability" was first conceptualized and mentioned in relation to pedagogy, there has been a "persistent humanism" which has led to the concept and pertinent discourse being viewed and discussed mainly from a human-centric perspective. While highlighting the persistent humanism, McKenzie and Bieler (2016) have stressed the need to rethink pedagogy in ways that embrace the emerging material conditions of the Anthropocene, and to see beyond the binaries of human and more-than-human entities. Nevertheless, the dominant discourse in sustainability pedagogy within ECEfS has been oriented towards educating the human/the child towards environmental stewardship, care and sympathy. In such a conceptualization, humans/children are seen as the protagonists in the phenomenon, and this perspective has led the notion of sustainability to centre mainly on the human, marginalizing the more-than-human constituents in the discourse.

Thus nurturing children with regard to environmental stewardship has been promoted in relation to curricular and pedagogical endeavours within ECEfS. Looking at it from post-humanism and/new materialism perspective, the notion of stewardship unintentionally entails "otherness" by describing the environment as something to be controlled, saved or mastered by human actors (Taylor 2013, 2017). This way of understanding the environment does not conform to the essence of post-humanism/new materialism, which stresses the mutually constitutive and entangled relationship between humans and the environment. Perhaps, as Reinertsen (2016) points out, it is worth shifting from stewardship to partnership since the latter has a potential to provoke consciousness about humans'/children's entanglement and their interconnections with the worlds—worldliness, as Haraway () calls it—in which they play out.

Hence, post-humanism/new materialism promotes pedagogy beyond the limits of developmentally appropriate practices and beyond the socialization of the individual child, and calls for early childhood pedagogy not to be confined by children's individual agency (Lenz Taguchi 2010; Blaise 2013; Taylor 2013). The approach instead calls for a relational child who is in constant entanglement with the heterogeneous non-human and more-than-human others, and argues that agency and individualism are not sufficient determinants of early childhood policy, curricula and pedagogy. As indicated by Lenz-Taguchi (2010), intra-active pedagogy can be a good tool to think with, as it highlights the intra-active relationships among living organisms and the materiality of the environment. Thus, early years educational structures and the learning environment of which the child becomes a part need to be designed in such a way that they can play an important role in strengthening children's relationship/connection to the more-than-human world.

The post-humanist/new materialist approach to curricula and pedagogy perceives children as living within an entangled common world where one cannot make a distinction between humans and more-than-human others. In this regard, post-humanism attempts to redefine/reposition the child in a common world (Taylor 2013) where he or she is entangled, related or connected to the physical world, the immediate environment and the place (Duhn 2012) that he/she shares with the more-than-human others in the immediate neighbourhoods, which can provide a rich pedagogical context.

Hence we (early childhood educators and researchers) need to reflect on our pedagogies and examine how we perceive children's relationship/entanglement with the world beyond humans and how that is manifested in everyday pedagogies in preschool settings. We need to question whether we are unintentionally being reductionist in our approach because we live a human-centric life and whether this might lead us to a disconnection from the world in which we live? I argue that this ontological standpoint promoted by post-humanism/new materialism has much to offer in terms of innovative ways of dealing with sustainability challenges to disrupt hegemonic/human-centric approaches and instead designing a pedagogical space that invites children to see the world in a wholly interrelated manner.

Moreover, unlike other educational levels, ECE typically has relatively open and flexible curricula, which allows the opportunity to design a learning environment that cultivates a holistic and relational world view, where humans and non-humans are living in entanglement, and hence educators can take advantage of this unique opportunity. In addition, children are imbued with curiosity and are open to new/different perspectives which paves the way to working with young children before they adopt an anthropocentric perspective and start to create boundaries which lead to false binary assumptions, such as human versus non-human and nature versus culture. I strongly argue that in a field like sustainability, where anthropocentrism is a significant challenge, post-humanism and new materialism align well with the ethos of sustainability and offer opportunities to rethink and redesign a pedagogical space that consciously recognizes the inevitable interdependence between human and non-human actors.

Parallel to curriculum and pedagogical issues, as thematically summarized in Table 1, earlier research inquiry within ECEfS tends to be human/child-centred. Further, to understand the underpinning ontological and epistemological beliefs and assumptions within ECEfS research, it is important to reflect on the theoretical/philosophical and methodological approaches/frameworks employed by prior research in the field. As described earlier, positivist, interpretivist, critical-theory and rights-based approaches have been identified as the most common theoretical approaches in the field, and they tend to address human characteristics, focusing on the child, its agency and relationship with nature. This again confirms that the research orientation in the field has been inherently human-centred and further strengthens the need for different, alternative and broader theoretical orientations. Besides theories, there is also a need for broader methodological perspectives that can overcome the methodological individualism that underlies many of the approaches to sustainability in early childhood, aiming towards more inclusive and alternative ways of understanding that are typically absent in ECEfS discourses.

However, this does not mean that there are no early childhood studies that attempt to use theories that explore issues beyond the human child. A review by Somerville and Williams (2015) highlighted a few studies that employed a post-human perspective. Among these are Duhn (2012) on pedagogy and place; Ritchie (2013) on sustainability and relationality; and Bone (2013) and Timmerman and Ostertag (2011) on animals and the environment. Joining these already existing initiatives, and following on from the recommendation by Somerville and Williams (2015) of the potential of post-humanism for researching planetary sustainability, I argue for the greater use of this perspective as an important theoretical/philosophical tool for rethinking methodologies and methods when addressing research, policy, pedagogy and curricular endeavours within ECEfS.

Concurrent with the argument presented here, several scholars have reacted to the dominant human exceptionalism view and have called for a change of perspectives and practices in dealing with sustainability challenges. Reinertsen (2016) points out that we should not be tied up with human action, but rather we should examine human and more-than-human relational assemblages of affecting and being affected. This urges us to rethink, reconstruct and deconstruct the notion of sustainability and its underlying discourse. Reinertsen (2016) states that it is time for us (humans) to open our eyes and ears to watch and listen to all the stories that non-humans are telling us. Likewise, Gibson et al. (2015) indicated that to reverse or change damaging human-centric behaviours, we first need to change our way of thinking and our ontological standpoint. They argue that if we are to see ourselves as part of the environment, not separate from it, and to reframe the environment in ethical terms, we must be able to learn from what is already happening in the world.

Arguing for a different ethical standpoint, Wals (2007) highlights that "we need nothing short of a new global ethic—an ethic which espouses attitudes and behaviour for individuals and societies which are consonant with humanity's place within the biosphere; which recognises and sensitively responds to the complex and ever-changing relationships between humanity and nature and between people" (p. 35). Likewise, Somerville (2015), while calling for an ontological move, points out the power of discipline and recommends a new disciplinary area called "ecological education". These are important proposals, indicating the need to create alternative knowledges and understandings towards a "sustainable" world. Building on these scholars' work, I emphasize the potential of the post-humanism/new materialism perspective as an important theoretical approach for challenging, reconstructing, deconstructing and expanding the notion of sustainability within ECE. The ECEfS field should not be inextricably tied to a human-centred, cognitive, meaning-making process while dealing with sustainability. Rather, the field must open up possibilities for new ways of being and becoming, to create "new" or alternative knowledge trajectories.

Although their work is not directly related to sustainability, early childhood researchers such as Lee (2002), Prout (2005) and Olsson (2009) have utilized the post-human perspective and offered outstanding scholarship in educational research. Scholars within ECEfS (Duhn 2012; Ritchie 2013; Somerville 2015) have also addressed the more-than-human world in different ways. Somerville and Green (2015) addressed sustainability in relation to place, while Duhn (2012) and Ritchie (2013) have dealt with indigenous and placed-based practices in Australia and New Zealand, respectively. These important contributions are signals indicating the emergence of this field within ECE in general and within ECEfS in particular. Building on the foundations laid by these scholars, this chapter has attempted to indicate how the post-humanism/new materialism perspective presents a comprehensive context which can offer alternative ways of looking at sustainability within ECEfS.

Some Concluding Remarks

This chapter has traced the idea of ECEfS from its historical roots, and through international dialogue, curricular frameworks, pedagogical practices and research perspectives. Arising from the literature reviewed here, two major features of ECEfS have been identified which shed light on the dominant discourse and characteristics of the field. The first is its deep-rooted and inherent human-centric/child-centric characteristic and its strong reliance on children's agency. The second is its emphasis on the environmental aspect of sustainability and its deterministic approach, which is aimed at nurturing stewardship, care and sympathy in young children. Drawing on the post-humanism/new materialism framework, I have attempted to deconstruct, reconstruct and expand the notion of sustainability so that it can be viewed beyond anthropocentric limitations, which could possibly lead to different/alternative forms of subjectivity and agency. In doing so, I pose fundamental questions about the ontological, epistemological and ethical starting points and suggest a rethinking of the idea of sustainability within ECEfS.

The relatively open nature of ECE makes it a field that is well situated to adapt to post-humanism and new materialism approaches since these perspectives allow us to see learning in a holistic and relational way without reducing it to binaries and hierarchies. These theories help us to have a unified world view which emphasizes humans' entangled relationship with the more-than-human world instead of breaking down learning into particular domains and dichotomies. Apart from relying on children's agency, I argue that post humanism and/new-materialism framework have the potential to creates a comprehensive context for conceiving a broader, complex and interconnected world (inhabited by human and non-humans) which could help us understand "sustainability" from a relational perspective.

Hence, ECE should not be content with the notion of children's agency and their cognitive ability to actively participate in issues pertaining to sustainability. The notion of children's agency is undoubtedly important but its adequacy for creating condition for understanding and engaging in a range of complex issues such as those related to sustainability has to be further examined. Thus I argue that employing these contemporary ways of thinking (post-humanism/new materialism) offers a broader perspective and leads to different ways of stretching boundaries and generating alternative discourses within ECEfS which can help us to understand children together with the non-human world that they are entangled with.

Finally, as a human reader, you might wonder about the inevitability of being human and the challenge to decentre oneself and pay attention to non-human/more-than-human others. Post-humanism and new materialism are not aimed at the victory of non-humans over humans. In the context of this chapter, they can instead be understood as remedial efforts to redress today's anthropocentric conditions (which contributes for "our" unsustainable life) by decentring humans sufficiently to recognize other, hitherto neglected and marginalized, more-than-human actors and their entangled relationship with humans, and the significance embedded in the relationship.

The Way Forward

Following the call for the rethinking of human relations with the environment and the more-than-human world by Taylor (2013); Reinertsen, (2016); Gibson et al. (2015) and Somerville (2015), I have attempted to elucidate how posthumanist/new materialist perspectives can help us to critically question assumptions at play within ECEfS and stretch the notion of sustainability itself by disrupting the deep-rooted dominant child-centric narrative.

Although early childhood researchers such as Lee (2002), Prout (2005), Taylor (2013), Rautio and Jokinen (2015), Lenz-Taguchi (2010), Blaise (2013) and Pacini-Ketchabaw & Nxumalo (2015) employ posthumanism and/new-materialism approaches for researching various aspects of children's life, they have not been well utilised for addressing sustainability challenges and hence more empirical studies are needed to indicate how these approaches can be implemented within ECEfS research and pedagogical activities. Thus, early childhood researchers and educators working on sustainability are encouraged to employ these theories as a tool to reinvestigate and how they plays out in the lifeworld of children and the way to embrace them within sustainability pedagogy in preschool settings.

Instead of providing answers to all the queries mentioned earlier, this chapter attempts to raise provocative questions that challenge contemporary (anthropocentric) understandings of sustainability. How do these different ontological, epistemological and ethical understandings shape "our" understanding and inform practices within the field of sustainability education in general and ECEfS in particular? What is actually being sustained in "sustainability" education? Who does the sustaining? Are human actors sufficient to deal with the challenge? Do we need other actors beyond human? How do posthumanism and/new materialism "redefine" the child and offer alternative view? Reflecting on these questions has a potential to generate different ways of looking at and dealing with sustainability challenges within ECEfS and beyond.

Moreover, early childhood researchers and educators need to closely examine policy discourses, curricular documents, pedagogical philosophies and research orientations that might unintentionally depict false dichotomies such as human-nonhuman, subject-object, nature-culture and the like. Given their potential to move the field of sustainability, it is worth inquiring how empirical research within ECEfS can be conducted within a posthumanist/new materialist research paradigm. To this effect, this chapter is an effort to deconstruct and then build different/alternative meanings for contemporary notions of sustainability.

These contemporary approaches offer alternative world views, and provide different methodological approaches and innovative analytical frameworks that can help us to see the complex interconnectedness in the world we live, and that in turn would create a comprehensive context to deal with sustainability challenges in early childhood education and beyond.

References

Ärlemalm-Hagsér, E. (2013). Respect for nature-a prescription for developing environmental awareness in preschool. *CEPS Journal: Center for Educational Policy Studies Journal, 3*(1), 25.

Ärlemalm-Hagsér, E. (2014). Participation as "taking part in": Education for sustainability in Swedish preschools. *Global Studies of Childhood, 4*(2), 101–114.

Barad, K. (2007). *Meeting the universe halfway: Quantum physics and the entanglement of matter and meaning*. Durham: Duke University Press.

Barratt Hacking, E., et al. (2007). Engaging children: Research issues around participation and environmental learning. *Environmental Education Research, 13*(14), 529–544.

Barratt, R., Barratt-Hacking, E., & Black, P. (2014). Innovative approaches to early childhood education for sustainability in England. In J. Davis & S. Elliot (Eds.), *Research in early childhood education for sustainability: International perspectives and provocations* (pp. 225–247). London: Routledge.

Bennett, J. (2010). *Vibrant matter: A political ecology of things.* Durham: Duke University Press.

Blaise, M. (2013). Activating micropolitical practices in the early years. In R. Coleman & J. Ringrose (Eds.), Deleuze and research methodologies (Deleuze connections series). Edinburgh: Edinburgh University Press.

Bone, J. (2013). The animal as fourth educator: A literature review of animals and young children in pedagogical relationships. *Australasian Journal of Early Childhood, 38*(2), 57–64.

Braidotti, R. (2013). *The posthuman.* Cambridge: Polity Press.

Brundtland, G. H. (1987). *Report of the World Commission on environment and development: Our common future.* Geneva: United Nations.

Chawla, L., & Rivkin, M. (2014). Early childhood education for sustainability in the United States of America. In J. Davis & S. Elliot (Eds.), *Research in early childhood education for sustainability: International perspectives and provocations* (pp. 248–265). London: Routledge.

Davis, J. (2009). Revealing the research 'hole'of early childhood education for sustainability: A preliminary survey of the literature. *Environmental Education Research, 15*(2), 227–241.

Davis, J., & Elliott, S. (2014). *Research in early childhood education for sustainability: International perspectives and provocations.* London: Routledge.

DeLanda, M. (2008). Deleuze, materialism and politics. In I. Buchanan (Ed.), *Deleuze and politics* (pp. 160–177). Edinburgh: Edinburgh University Press.

Deleuze, G., & Guattari, F. (1987). *A thousand plateaus: Capitalism and schizophrenia.* London: Continuum.

Dolphijn, R., & Van der Tuin, I. (2012). *New materialism: Interviews & cartographies.* Ann Arbor: Open Humanities Press.

Duhn, I. (2012). Places for pedagogies, pedagogies for places. *Contemporary Issues in Early Childhood, 13*(2), 99–107.

Edwards, S., & Cutter-Mackenzie, A. (2011). Environmentalising early childhood education curriculum through pedagogies of play. *Australasian Journal of Early Childhood, 36*(1), 51–59.

Engdahl, I., & Ärlemalm-Hagsér, E. (2014). Education for sustainability in Swedish preschools. In J. Davis & S. Elliot (Eds.), *Research in early childhood education for sustainability: International perspectives and provocations* (pp. 208–224). London: Routledge.

Engdahl, I., & Rabusicova, M. (2011). Children's voices about the state of the earth. *International Journal of Early Childhood, 43,* 153–176.

Gambino, A., Davis, J., & Rowntree, N. (2009). Young children learning for the environment: Researching a forest adventure. *Australian Journal of Environmental Education, 25*, 83–94.

Gibson, K., Rose, D. B., & Fincher, R. (2015). *Manifesto for living in the anthropocene*. Brooklyn: Punctum.

Gilbert, L., Fuller, M., Palmer, S., & Rose, J. (2014). Early childhood education for sustainability in the United Kingdom. In J. Davis & S. Elliot (Eds.), *Research in early childhood education for sustainability: International perspectives and provocations* (pp. 294–308). London: Routledge.

Hadzigeorgiou, Y., Prevezanou, B., Kabouropoulou, M., et al. (2011). Teaching about the importance of trees: A study with young children. *Environmental Education Research, 17*(4), 519–536.

Haraway, D. (2008). *When species meet*. Minneapolis: University of Minnesota Press.

Hedefalk, M., et al. (2014). Education for sustainable development in early childhood education: A review of the research literature. *Environmental Education, 21*(7), 975–990.

Ji, O., & Stuhmcke, S. (2014). The project approach in early childhood education for sustainability. In J. Davis & S. Elliot (Eds.), *Research in early childhood education for sustainability: International perspectives and provocations* (p. 158). London: Routledge.

Kahriman-Öztürk, D., Olgan, R., & Güler, T. (2012). Preschool children's ideas on sustainable development: How preschool children perceive three pillars of sustainability with the regard to 7R. *Educational Sciences: Theory and Practice, 12*(4), 2987–2995.

Latour, B. (2004). *The politics of nature: How to bring science into democracy*. Cambridge, MA: Harvard University Press.

Latour, B. (2005). *Reassembling the social: An introduction to actor network theory*. Oxford: Oxford University Press.

Lave, J., & Wenger, E. (1991). *Situated learning: Legitimate peripheral participation*. Cambridge: Cambridge University Press.

Lee, N. (2002). *Childhood and society: Growing up in an age of uncertainty*. Buckingham: Open University Press.

Lenz-Taguchi, H. (2010). *Going beyond the theory/practice divide in early childhood education: Introducing an intra-active pedagogy*. London: Routledge.

Lenz-Taguchi, H. (2014). New materialism and play. In E. Brooker et al. (Eds.), *Sage handbook of play and learning in early childhood*. California: Sage.

Mackey, G. (2014). Valuing agency in young children: Teachers rising to the challenge of sustainability in the Aotearoa New Zealand early childhood context. In J. Davis & S. Elliot (Eds.), *Research in early childhood education for sustainability: International perspectives and provocations* (pp. 180–193). London: Routledge.

Malone, K., Truong, S., & Gray, T. (2017). *Reimagining sustainability in precarious times*. Singapore: Springer.
McKenzie, M., & Bieler, A. (2016). *Critical education and sociomaterial practice: Narration, place and the social*. Bern: Peter Lang Press.
Mickey, S. (2007). Contributions to anthropocosmic environmental ethics. *Worldviews: Global Religions, Culture, and Ecology, 11*(2), 226–247.
Mickey, S. (2010). Ecology in Deleuze and Guattari. In S. Esbjörn-Hargens (Ed.), *Integral theory in action: Applied, theoretical, and constructive perspectives on the AQAL model* (p. 325). New York: SUNY Press.
Mickey, S. (2016). *Coexistentialism and the unbearable intimacy of ecological emergency*. Maryland Rowman & Littlefield.
Miller, M. G. (2014). Intercultural dialogues in early childhood education for sustainability. In J. Davis & S. Elliot (Eds.), *Research in early childhood education for sustainability: International perspectives and provocations* (pp. 63–78). London: Routledge.
O'Gorman, L. M. (2014). The arts and education for sustainability: Shaping student teachers' identities towards sustainability. In J. Davis & S. Elliot (Eds.), *Research in early childhood education for sustainability: International perspectives and provocations* (pp. 266–279). London: Routledge.
O'Gorman, L., & Davis, J. (2012). Ecological footprinting: Its potential as a tool for change in Preservice teacher education. *Environmental Education Research, 19*(6), 779–791.
Olsson, L. M. (2009). *Movement and experimentation in young children's learning: Deleuze and Guattari in early childhood education*. London: Routledge.
Pacini-Ketchabaw, V., & Nxumalo, F. (2015). Unruly raccoons and troubled educators: Nature/culture divides in a childcare center. *Environmental Humanities, 7*, 151–168.
Phillips, L. (2014). I want to do real things. In J. Davis & S. Elliot (Eds.), *Research in early childhood education for sustainability: International perspectives and provocations* (pp. 194–207). London: Routledge.
Piaget, J. (1997). Development and learning. In M. Gauvain & M. Cole (Eds.), *Readings of the development of children* (pp. 19–28). New York: W.H. Freeman and Company.
Poe, A. (2011). Diana Coole and Samantha Frost (eds), New materialisms: Ontology, agency, and politics. *Journal of French and Francophone Philosophy, 19*(1), 153–164.
Pramling Samuelsson, I., & Kaga, Y. (red.). (2008). *The contribution of early childhood education to a sustainable society*. Paris: UNESCO.
Prout, A. (2005). *The future of childhood*. Abingdon: New York Routledge Falmer.
Rautio, P., & Jokinen, P. (2015). Children's relations to the more-than-human world beyond developmental views. In T. Skelton, J. Horton, & B. Evans (Eds.), *Geographies of children and young people: Play, recreation, health and well being*. Singapore: Springer.

Reinertsen, A. B. (2016). Becoming Earth. In *Becoming Earth* (pp. 1–13). Rotterdam: Sense Publishers.

Ritchie, J. (2013). Sustainability and relationality within early childhood care and education settings in Aotearoa New Zealand. *International Journal of Early Childhood, 45*(3), 307–326.

Siraj-Blatchford, J., Smith, K. C., & Pramling Samuelsson, I. (2010). *Education for sustainable development in the early years.* Gothenburg: OMEP.

Somerville, M. (2015). Environment and sustainability education: A fragile history of the present. In D. Wyse et al. (Eds.), *The SAGE handbook of curriculum, pedagogy and assessment.* London: Sage.

Somerville, M., & Green, M. (2015). *Children, place and sustainability.* Basingstoke: Palgrave MacMillan.

Somerville, M., & Williams, C. (2015). Sustainability education in early childhood: An updated review of research in the field. *Contemporary Issues in Early Childhood, 16*(2), 102–117.

Sundberg, B., & Ottander, C. (2014). Science in preschool–a foundation for education for sustainability? In J. Davis & S. Elliot (Eds.), *Research in early childhood education for sustainability: International perspectives and provocations* (p. 280). London: Routledge.

Taylor, A. (2013). *Reconfiguring the natures of childhood.* London: Routledge.

Taylor, A. (2017). Romancing or reconfiguring nature in the Anthropocene? Towards common worlding pedagogies'. In K. Malone, T. Gray, & S. Truong (Eds.), *Reimagining sustainability education in precarious times* (pp. 61–75). Amsterdam: Springer.

Taylor, C., & Hughes, C. (2015). *Posthuman research practices in education.* London: Palgrave MacMillan.

Timmerman, N., & Ostertag, J. (2011). Too many monkeys jumping in their heads: Animal lessons within young children's media. *Canadian Journal of Environmental Education, 16*, 59–75.

UNESCO. (2005). *Decade of education for sustainable development 2005–2014. Draft international implementation scheme.* Paris: UNESCO.

United Nations. (1992). *Agenda 21: The United Nations programme of action from Rio.* New York: United Nations.

United Nations Millennium Summit. (2015). Retrieved September 10, 2016, from http://www.un.org/en/events/pastevents/millennium_summit.shtml

Vygotzky, L. S. (1986). *Thought and language.* Cambridge, MA: Harvard University Press.

Wals, A. E. J. (2007). Think piece. Learning in changing world and changing in a learning world: Reflexively fumbling towards sustainability. *Southern African Journal of Environmental Education, 24*, 35–45.

Wals, A. E. J. (2012). Learning our way out of un-sustainability: The role of environmental education. In S. Clayton (Ed.), *Handbook on environmental and conservation psychology* (pp. 628–644). Oxford: Oxford University Press.

Wolfe, C. (2010). *What is posthumanism?* (Vol. 8). Minneapolis: University of Minnesota Press.

Young, T., & Cutter-MacKenzie, A. (2014). An AuSSI early childhood adventure: Early childhood educators and researchers actioning change. In J. Davis & S. Elliot (Eds.), *Research in early childhood education for sustainability: International perspectives and provocations* (pp. 143–157). London: Routledge.

Windows on a Changing World: Using Children's Literature as an Aesth/Ethical Trope in Early Years Education for Sustainability

Dawn L. Sanders

Introduction

This chapter is an exploration of the possibilities afforded children's imaginations by three books. It constitutes a teacher's situated reflection on how the heterotopic "nowhere and here" of children's literature can be a metaphorical window to "something or somewhere else" in the context of environmental sustainability. Children's stories are reflected on as points of departure from which to consider an aesth/ethical (Bergmann 2005) trope between humans, more-than-humans and material matter. The chosen texts are used in contemporary teaching to inspire becoming teachers' reflections in early childhood courses, thus they are "social objects" (McGann 2005) in an environmental learning context. Moreover, each text is relevant to urban and suburban life, thereby providing a literary counterpoint to stories that "ascribe sensory experience of 'real' nature as the only true source for environmental commitment" (Garrison et al. 2015, p. 188). Each of the books in focus utilizes windows in different narrative contexts: in Björk and Anderson (1978) the window is a physical

D.L. Sanders (✉)
Department of Pedagogical, Curricular and Professional Studies,
University of Gothenburg, Gothenburg, Sweden

O. Franck, C. Osbeck (eds.), *Ethical Literacies and Education for Sustainable Development*, DOI 10.1007/978-3-319-49010-6_8

place around which a garden evolves, whereas in Baker (1991, 2002) the window is a metaphor for change, while Light (2006) uses the window as a symbol of home that has been reframed in a dystopian world devoid of green.

> by opening a window in our minds, by understanding how change takes place and by changing the way we personally affect the environment, we can make a difference. (Baker 2002, Author's note)

Whitin and Whitin (1996) state that "the real sustaining force of inquiry learning lies at the window because that is where learners meet the world face to face" (p. 87). To examine their claim that "the sustaining force of inquiry learning lies at the window", we need first to consider the physical phenomenon of the window itself, and its sociomaterial function between " home" and "the wider world". For example, how do we distinguish children's experiences of the view *through* the window in contrast with their experiences of different phenomena *at* the window? In seeing the green tree in the distance, does a young child learn what it means to be green or indeed something about what it means to be a tree? In pressing their nose close to the glass, does a child physically feel the window, its temperature and material density? Is the child attending to people walking past the tree or the changes occurring to their own nose? Can a child's "frame of attention" (Marton 2015) concerning the window be transformed through literary settings? How might the variations between being *at* the window and looking *through* a metaphorical window impact on children's "entwined worlds" of "elsewhere" (Taylor 2013)?

The argument presented here subscribes to the dialogical possibilities that fictional literature can create, both in and beyond the present. Hence Light (2006) offers children the possibility of imagining a future without flowers and feeling what that might be like. Baker (1991/2002) provides a metaphor through which to examine environmental change by crossing temporal zones between past, present and future, and in so doing she gives children a prompt to ask "what if" questions. Meanwhile, Björk and Anderson (1978) mix instructional and fictional genres to present a story in which the reader both witnesses the growing of plants and is informed how to grow one, thus providing opportunities to make a windowsill garden in the here and now of children's lives.

Three children's books inform this chapter: one each from Sweden, Australia and the UK: Christina Björk and Lena Anderson's *Linnea's windowsill garden* (1978 [*Linnea planterar*]), Jeannie Baker's *Window* (1991/2002) and John Light's *The flower* (2006). Two focus on the act of growing plants from seed: Björk and Anderson's book makes border crossings between fictional and non-fictional genres in an intergenerational story of a girl and her neighbour, Mr Bloom, cultivating a windowsill garden in their apartment block, while Light offers a grey monochrome world in which growing flowers has disappeared from social practice, thus cultivating seeds is represented as an act of transgression.

All three books use visual images as narrative constructs. Indeed, Baker's story relies solely on an image-based narrative. Furthermore, all of the texts afford discourse-rich responses oriented around plant life, environmental change and social interaction. The publication period of the selected books spans a key tipping point in the earth's history when environmental change, as a result of human activities, has accelerated—a period now called the Anthropocene (Steffen et al. 2007). The books remain in circulation despite their original publication dates, demonstrating their sustained value in contemporary children's literature. Indeed, a new edition of *Window* was published in 2002.

The books in focus can be seen to represent specific "turns" in children's environmental literature and consequently can be interpreted in relation to wider discussions about fiction and environmentalism, such as those concerning Cormac McCarthy's novel *The road* (2006). The environmentalist George Monbiot considers *The road* to be the most important environmental book ever written. He defends his choice by noting that McCarthy

> Exposes the one terrible fact to which our technological hubris binds us: our dependence on biological production remains absolute. Civilisation is just a russeting on the skin of the biosphere, never immune from being rubbed against the sleeve of environmental change. (Monbiot 2007, p. 294)

Building on Monbiot's metaphor, Holm (2012) suggests that the disaster narrative in *The road* "does not *produce* human vulnerability, it rather *reveals* an inherent vulnerability that was there all the time" (p. 234). This vulnerability to environmental change can also be found in children's fiction. In choosing the three books, I seek to provoke discourse on the use of literary forms of art as a reflexive tool in early years education. Each represents a particular genre in environmental story-making, so the focus is not one of compare and contrast but rather examples of books, which, in their various forms, "encourage young readers to see their worlds differently" (Taylor 2013, p. 73).

Small Worlds and the Big World

The represented genres include a dystopian view of plants in a dust-filled world, absent of flowers (Light 2006), a world in which a window bears witness to a cycle of deforestation and urbanization (Baker 1991/2002), and an intimate look at intergenerational gardening in an apartment (Björk and Anderson 1978). In recognition of recent post-humanist discussions in early childhood education (ECE; e.g. Taylor 2013), we can ask if activities such as making a windowsill garden (Bjork and Anderson 1978; Light 2006) merely offer human agency in giving opportunities to "shape small worlds", and thus provide knowledge and feelings that "they can participate in shaping the big world tomorrow" (Sobel 1990, p. 12). Do such presupposed continuums of knowledge, feelings and participation in "the big world" underproblematize inter- and intraspecies agency, empathetic capacity and the complex material relationships that children reconstruct in such "small worlds"? Jørgensen (2014) suggests that ECE models often affirm a "hierarchy of learning in nature" oriented around a recurrent incremental knowledge narrative of "child saving the world" (Jørgensen 2014, p. 27). In providing stories in which "children's lives are inseparably bound up with all manner of other lives, other forces and other things" (Taylor 2013, p. 79), the three texts can be seen to create an assemblage of "nowhere and here" (Taylor 2009) in which to find multiple and relational nodes of ethics concerning agency, sense of place and materiality (Kronlid and Öhman 2013).

Both "Nowhere and Here"

Murdoch (1977) considers art "to be far and away the most educational thing we have" (p. 86), and Posner (1997) has suggested that "immersion in literature" can make us "better citizens" (p. 2) in the moral sense of Nussbaum's "sympathetic imagination" (1995). Such statements retain a powerful call to engage with art and literature in the context of critically responsive education. However, underlying such calls lies a "persistent humanism" (McKenzie and Bieler 2016, p. XIII), which could be seen to restrict the ethical possibilities of these expressive forms to a human-centric domain. A further challenge to the use of fiction as an aesth/ethical (Bergmann 2005) trope is that "literature's space is deeply ambiguous: nowhere and here" (Johnson 2006, p. 86), prompting questions as to how literature might contribute to an environmental education embedded in "sense of place". Andrews (2014) suggests that "through narrative imagination we are both anchored *and* transported" (p. 2, my italics), indicating that the ambiguity provided by a literary "nowhere and here" might be a productive contribution to environmental learning. Certainly the contribution of ambiguity, as opposed to vagueness, in an arts-based education for sustainability has been affirmed in a recent study by Eernstman and Wals (2013). Moreover, "when Ricoeur locates the 'productive imagination' in *fiction*, in the 'nowhere' that fiction provides, the paradox is that fiction provides a new dimension of reality" (Taylor 2009, p. 98). Therefore I would argue that such "augmentation of reality" (Ricoeur, cited in Taylor 2009), through the "nowhere and here" of fiction, affords new possibilities for opening up conversations with young children about the social and environmental ethics of place and the more-than-human world, and questions of fragility.

In exploring Foucault's definition of heterotopia, Johnson (2006) states that it is "crucial" to reference "that which draws us out of ourselves" (p. 84). He goes on to describe such perspectives:

> Heterotopias draw us out of ourselves in peculiar ways; they display and inaugurate a difference and challenge the space in which we may feel at home. These emplacements exist out of step and meddle with our sense of interiority. (Johnson 2006 p. 84)

Thus if we apply these Foucauldian notions of heterotopia to the three texts, we can consider ways to "splinter the familiar" (Johnson 2006 p. 85) configuration of child and nature through the windows of "home". It is here that a heterotopic lens offers, as Johnson observes, "different degrees of relational intensity" and hence can provide illumination for Ricoeur's "productive imagination" (Taylor 2009), and the potential for both value-oriented and relation-oriented discourses (Kronlid and Öhman 2013) situated between the text and the reader. In addition, the diverse genres represented by the three stories afford the reader the possibility of a "disclosure of reality that is both available and yet to come" (Taylor 2009, p. 99). So, for example, *Linnea's windowsill garden* is grounded "in the available" through an instructional text contained within a fictional story of an intergenerational meeting between a child (Linnea) and an adult (Mr Bloom; Fig. 1), central to which is the sharing of gardening knowledge and practice, whereas *Window* presents both "the available and the yet to come" (ibid., p. 99) through a richly detailed visual narrative of change over time seen through 13 window collages.

Fig. 1 Linnea's Garden/Linnea Planterar. Reproduced by permission of Christina Björk and Lena Anderson

Finally, *The flower* is a children's picture book with large illustrations accompanied by short pieces of text. In essence, it presents "the yet to come" (ibid.) of a dystopian view of an imagined future without soil or flowering plants. Books about growing flowers are labelled "dangerous", and obtaining seeds is presented as an act of transgression to be kept secret. Light's story presents soil as an extinct material, which "Brigg", the main character, has to re-form from collected dust "from all over the city ... until he had enough to fill a mug" (Light 2006, unnumbered page; Fig. 2). On the final page, where Brigg is shown looking out of his window framed by flowering plants, there are no words, so the reader is left wondering if this is his future or if it is Brigg's imagining of a possible future. In contrast, Linnea's windowsill garden appears to be an image of fecundity, albeit one that is physically and socially restrained by the limitations of urban apartment living. It is no coincidence that each book reflects historical aspects of environmental discourse—the dystopian visions of twenty-first century environmentalism as embodied in Monbiot's review of *The road* and the utopian dreams of twentieth-century self-sufficiency. Situated in the middle of the publication timeline, Baker's original edition in its visual

Fig. 2 *The flower*. Reproduced by kind permission of Child's Play (International) Ltd. Text © John Light 2006. Illustrations © Child's Play (International) Ltd 2006. First published in 2006 by Child's Play (International) Ltd. All rights reserved

orientation creates spaces for a range of interpretations mirroring post-modern approaches to literary forms.

Returning to the window as a place where "learners meet the world face to face" (Whitin and Whitin 1996), we can now reinscribe the experience of planting seeds (Björk and Anderson 1978; Light 2006), the handling of soil (Björk and Anderson 1978; Light 2006), being with and talking to Mr Bloom (Björk and Anderson 1978), witnessing environmental change (Baker 1991), and watching plants grow and flowers open (Björk and Anderson 1978; Light 2006) in a monochrome urbanized world (Light 2006) as an opportunity to "live responsively" in "questioning relationships" (Haraway 2008) across species, generations and materiality. If the three texts in question are conceptualized as heterotopic lenses through a fictional window where "things are different", might we then reimagine children's stories as occupying a heterogeneous space in which various readings of place and relationships can be drawn out between child and adult that recognizes, and makes public, "the tensions, contradictions, uncertainties and ambiguities of constructing an ecological identity" (Thomashow 1998, p. xv) in the post-modern world, with all the ethical complexity that such an identity brings. Moreover, is this a space in which Ricoeur's "productive imagination" might enable transformations between children and their environment, thus engaging their "sympathetic imagination" (Nussbaum 1995)? In threading together story, place and activity, might these narratives and other children's fiction assist us, as educators, to reconstruct new environmental roles for "story-time" and "the home-corner" in ECE settings?

Something or Somewhere Else

The journey to "something or somewhere else" (Radford et al. 2015, p. 743) has the potential to awaken children's "sympathetic imagination" (Nussbaum 1995) in, and beyond, the human world. Moreover, Ricoeur's demonstration that "imagination is not something marginal to or occasional in thought but rather permeates all thought and conceptualization" (Taylor 2009, p. 94) is, I believe, a powerful construct for bridging the aesth/ethical gap (Bergmann 2005) between "the marks we leave on a fallen world" and "what kinds of marks we wish to leave" (Cronon 1995, p. 18). Thus the fictional literature chosen by teachers for early childhood settings is a significant act in relation to young children's emergent ecological identities and their capacity for ethical thinking

(Burke and Cutter-Mackenzie 2010). Moreover, these situated reflections indicate wider implications for the role of curriculum and the specification of fictional texts; a place in which there are sometimes contested territories between governmental policy, teacher training and teacher practice. In framing young children's literature in this way, I advocate for fiction as "no mere frill" (Nussbaum 1995, p. 2) to human existence and consider its metaphorical abilities to be essential for meetings between humans and other species.

References

Andrews, M. (2014). *Narrative imagination and everyday life.* Oxford: Oxford University Press.

Baker, J. (1991/2002). *Window.* London: Walker Books.

Bergmann, S. (2005). Space and spirit: Towards a theology of inhabitation. In S. Bergmann (Ed.), *Architecture, aesth/ethics and religion* (pp. 45–103). Frankfurt am Main: IKO-Verlag für interkulturelle Kommunikation.

Björk, C., & Anderson, L. (1978). *Linnea's windowsill garden [Linnea planterar].* Stockholm: R&S Books.

Burke, G., & Cutter-Mackenzie, A. (2010). What's there, what if, what then, and what can we do? An immersive and embodied experience of environment and place through children's literature. *Environmental Education Research, 16*(3–4), 311–330.

Cronon, W. (1995). The trouble with wilderness; or, getting back to the wrong nature. In W. Cronon (Ed.), *Uncommon ground: Rethinking the human place in nature* (pp. 69–90). New York: W. W. Norton & Co.

Eernstman, N., & Wals, A. E. (2013). Locative meaning-making: An arts-based approach to learning for sustainable development. *Sustainability, 5,* 1645–1660.

Garrison, J., Östman, L., & Håkansson, M. (2015). The creative use of companion values in environmental education and education for sustainable development: Exploring the educative moment. *Environmental Education Research, 21*(2), 183–204.

Haraway, D. (2008). *When species meet.* Minneapolis/London: University of Minnesota Press.

Holm, I. W. (2012). The frailty of everything-Cormac McCarthy's *The road* and modern disaster discourse. In C. Meiner & K. Veel (Eds.), *The cultural life of catastrophes and crises.* Berlin/Boston: de Gruyter.

Johnson, P. (2006). Unravelling Foucault's "different spaces". *History of the Human Sciences, 19*(4), 75–90. doi:10.1177/0952695106069669.

Jörgensen, K. A. (2014). What does it mean for children's experiences when the kindergarten is moving their everyday activities outdoors? *Gothenburg Studies in Educational Sciences, 362.*

Kronlid, D. O., & Öhman, J. (2013). An environmental conceptual framework for research on sustainability and environmental education. *Environmental Education Research, 19*(1), 21–44.

Light, J. (2006). *The flower*. Swindon: Child's Play.

Marton, F. (2015). *The necessary conditions of learning*. London/New York: Routledge.

McCarthy, C. (2006). *The road*. New York: Alfred A Knopf.

McGann, J. (2005). From text to work. Digital tools and the emergence of the social text. In A. M. Hensen, R. Ludeke, W. Streit, C. Urchueguia & P. Shillingsburg (Eds.), *Variants 4: The book as artifact text and border* (pp. 225–240). Amsterdam/New York: The Journal of the European Society for Textual Scholarship.

McKenzie, M., & Bieler, A. (2016). *Critical education and sociomaterial practice: Narration, place and the social*. Bern: Peter Lang Press.

Monbiot, G. (2007) Civilisation ends with a shutdown of human concern. Are we there already? *The Guardian*, October 30. https://www.theguardian.com/commentisfree/2007/oct/30/comment.books. Accessed 20 June 2016.

Murdoch, I. (1977). *The fire and the sun: Why plato banished the artists*. Oxford: Oxford University Press.

Nussbaum, M. C. (1995). *Poetic justice: The literary imagination and public life*. Boston: Beacon Press.

Posner, R. A. (1997). Against ethical criticism. *Philosophy and Literature, 21*(1), 2.

Radford, G. P., Radford, M. L., & Lingel, J. (2015). The library as heterotopia: Michel Foucault and the experience of library space. *Journal of Documentation, 71*(4), 733–751.

Sobel, D. (1990). A place in the world: Adult memories of childhood's special places. *Children's Environments Quarterly, 7*(4), 5–13.

Steffen, W., Crutzen, P. J., & McNeill, J. R. (2007). The anthropocene: Are humans now overwhelming the great forces of nature? *Ambio, 36*(8), 614–621.

Taylor, G. H. (2009). Ricoeur's philosophy of imagination. *Journal of French Philosophy, 16*, 93–104.

Taylor, A. (2013). *Reconfiguring the natures of childhood*. New York/London: Routledge.

Thomashow, M. (1998) Ecological identity: becoming a reflective environmentalist MIT Press: Cambridge,

Whitin, D., & Whitin, P. (1996). Inquiry at the window. *Language Arts, 73*, 82–87.

Aesthetic Experiences Related to Living Plants: A Starting Point in Framing Humans' Relationship with Nature?

Eva Nyberg

Introduction

This chapter highlights some theoretical and empirical research regarding aesthetic experiences in science teaching and learning, and the significance that these experiences might have for developing environmental ethics. The Deweyan perspective of experience and continuity (e.g. Dewey, 1938/1997) is used as a theoretical lens and empirical data from educational situations illustrate the theoretical discussion. In these examples, teachers and student teachers are confronted by and given tools for the close study of plants—that is, when the teacher in educative situations explicitly makes the students observe and/or take care of plants.

After 25 years of science teaching, in upper-secondary school and in teacher education, as well as in professional development courses, I have gradually come to the insight that I have from the start, in different ways, tried to engage my students in science through a variety of sensory experiences. I have done this because of my belief that sensory experiences are key to interest, motivation, understanding and engagement regarding

E. Nyberg (✉)
Department of Pedagogical, Curricular and Professional Studies,
University of Gothenburg, Gothenburg, Sweden

O. Franck, C. Osbeck (eds.), *Ethical Literacies and Education for Sustainable Development*, DOI 10.1007/978-3-319-49010-6_9

both the science content and the students' perceptions of their interdependence with nature. Preparing a lecture some time ago on this theme, I suddenly remembered a statement in the preface of my thesis, seven years earlier, and included it in my script:

> As a high school teacher one of my pillars was to try to teach in such a way that the students would be fascinated and amazed by the beauty and richness of the natural environment, and gain an understanding of the way in which we are dependent on a functioning natural environment. I brought nature into the classroom in the form of litter or living plants and insects, let students use stereo microscopes to make them see how beautiful a piece of lichen, or how cool a beetle can look magnified and with additional light, or discover the beauty of diatoms, which in the microscope are like shimmering green jewelry. (Nyberg 2008, p. 13, translated from Swedish)

Recently I received a copy of Carol Rodgers' (2002) "Defining reflection: Another look at John Dewey and reflective thinking" from a colleague, and thereafter I returned to Dewey's writing with a different perspective from what I had during my doctoral study. I recognized that my philosophy of teaching, ideas of learning and previous research studies have more in common with what he describes than I had previously perceived. I therefore decided to further explore theoretically the role that aesthetic[1] experiences have in science teaching and learning, and the significance that these experiences might have for environmental ethics. One "critical incident" (Tripp 2012) in pushing this idea forward was a seminar with a group of science and technology teachers. I was focused on asking them to look at mosses and blueberry flowers in stereomicroscopes with the intent of highlighting the effect of magnification and additional light, and the representative from the Swedish National Agency for Education afterwards exclaimed, "You are right, we have forgotten this!" Another such event, which had made me reflect on my teaching practice, was when I transcribed student discussions on the lifecycles of the blueberry plant and it turned out that the students' exploration of the plants contained not only the biological concepts and explanations, which were asked for, but also a rich number of aesthetic discoveries and expressions.

Previous research I have been involved in has demonstrated that taking care of, and observing, living organisms in the classroom can elicit emotional feelings and have a positive impact on student interest both in science and in organisms that they previously considered uninteresting

(Nyberg 2004, 2008; Nyberg and Sanders 2014). The organisms in question were living maggot larvae and pea plants, which the students took care of and observed during the larvae's development to adult flies (ten days) and from seeds until new peas were formed (8–10 weeks). However, such impacts appear to be time limited (Nyberg 2008; Nyberg and Sanders 2014). Therefore, to make students build prolonged relationships with living organisms and to understand, for example, the role that plants play in life on earth and in human nutrition is complex and challenging. Nevertheless, it seems as if, with regard to plants, students attend more closely to plants, and observe and document them, when caring for individual specimens. Likewise, if the educator draws attention to specific details and characteristics of plants, the students might then "be more connected to the 'green side of life' " (Nyberg and Sanders 2014, p. 152).

Regarding the relevance of learning about basic ecological prerequisites for human life and activities through encounters with nature, Öhman and Sandell (2015) hold that such a learning approach can play an important role in education for sustainable development (ESD) and that "encounters with nature can both challenge and broaden motives and direction for sustainable development" (Öhman and Sandell 2015, p. 260, translated from Swedish). However, they conclude that although there is now increasing knowledge of how relations with nature are created through encounters with nature, we know "rather little about the continuity of this learning and to what extent experiences of encounters with nature constitute a resource in discussions about the environment and sustainability in other contexts" (Öhman and Sandell 2015, p. 264, translated from Swedish).

However, Garrison et al. (2015), in discussing values in educational settings, claim that if "we ascribe sensory experience of 'real' nature as the only true source for environmental commitment" (p. 188), any experience which does not involve direct contact with nature would be disqualified and hence, for example, the experience of all "those living in urban environments" (p. 188). "Direct contact" with nature can, however, be achieved through other means—that is, not necessarily implying "real" nature. These encounters can take place either outdoors or indoors, in a classroom or at home. Plants can be grown, living specimens can be brought indoors from a natural habitat nearby (and put back again), or larvae of different kinds can be purchased and set free when their development is completed. Therefore "direct encounters" with nature do not necessarily have to imply "real" nature. Nevertheless, in these encounters, when the students are given tools such as a stereoscope to discover details of specimens or a task to observe and take care of living organisms over time, my experience is that aesthetic as well as ethical dimensions have the potential to be an integral part of the science learning and eventually potential regarding building a sense of relationship with the natural world. The ethical dimension of both environmental education and ESD is, according to Kronlid and Öhman (2013), of increasing interest in both Swedish and international educational research, and the ethical responsibility concerns not only future generations but also nature in its own right (Öhman 2006; Östman 2015).

Dewey's Concept of Experience

To experience is to learn, according to Dewey; "there is an intimate and necessary relation between the process of actual experience and education" (Dewey 1938/1997, p. 20), but not all experiences are educative. He asserts that "The belief that all genuine education comes about through experience does not mean that all experiences are genuinely or equally educative" (p. 25); it all depends on the "quality of experience" (p. 27). He states that "the central problem of an education based upon experience is to select the kind of present experiences that live fruitfully and creatively in subsequent experiences" (p. 28), and further that "An experience is always what it is because of a transaction taking place between an individual and what, at the time, constitutes his environment" (p. 43). The environment here is "whatever conditions interact with personal needs, desires, purposes, and capacities to create the experience which is at hand" (p. 44). Every experience is, according to Dewey, "a moving force. Its value can be judged only on the ground of what it moves toward and into" (p. 38).

Dewey's principle of continuity of experience "means that every experience both takes up something from those which have gone before and modifies in some way the quality of those which come after" (p. 35). Dewey writes: "As an individual passes from one situation to another, his world, his environment, expands or contracts ... What he has learned in the way of knowledge and skill in one situation becomes an instrument of understanding and dealing effectively with the situations which follow. The process goes on as life and learning continue" (p. 44). He also states that "Continuity and interaction in their active union with each other provide the measure of the educative significance and value of an experience. The immediate and direct concern of an educator is then with the situations in which interaction takes place" (p. 45). As expressed by

Rogers (2002), "Interaction and continuity, the elements of experience, are the *x* and *y* axes of experience. Without interaction learning is sterile and passive, never fundamentally changing the learner. Without continuity learning is random and disconnected, building toward nothing either within the learner or in the world" (p. 847). According to Dewey, the responsibility of educators is that "they should know how to utilize the surroundings, physical and social, that exist so as to extract from them all that they have to contribute to building up experiences that are worth while" (Dewey 1938/1997, p. 40).

Dewey's holistic concept of an experience implies that he objects to the separation between, for example, science and emotions, and he argues that there is no division between "science, morals, and esthetic appreciation" (Dewey, 1929, p. 407), stating that

> It is not possible to divide in a vital experience the practical, emotional and intellectual from one another and to set the properties of one over against the characteristics of the others. The emotional phase binds parts together into a single whole; "intellectual" simply names the fact that the experience has meaning; "practical" indicates that the organism is interacting with events and objects which surround it. (Dewey, 1934/1987, p. 61)

Aesthetics and Sensory Experiences in Science Teaching and Learning

In referring to Dewey's concept of continuity, Cobern et al. (1999) state that a person's experience is continuous. Yet "as children grow, and certainly as adults, many learn to box off portions of their thought lives so that, for example, scientific and aesthetic knowledge become separately and exclusively boxed" (pp. 542–543). Their interview study concerned 16 ninth graders' conceptualizations of nature. Significantly, during the interviews the interviewer never introduced science in the conversation. If it came up in conversation it was at the informant's initiative only. For most of the students the science they learned in school seemed to have little to do with the natural world they knew from personal experience. By contrast, most of the students "attached considerable importance to personal experiences with Nature" (p. 553) and their "environmental inclinations were strong" (p. 553). However, their narratives could not, according to the authors, be easily associated with anything related to school but instead disclosed a "disjunction between the students' experience of the world and the world as constructed in the science classroom" (p. 557). The authors compare this with Eger (1992), who calls this the "double distancing" (p. 342) between science and nature in the science classroom. Cobern et al. (1999) conclude that their research suggests that "science education does too little to help students integrate the important concepts of their own worlds with the important concepts of science" (p. 557).

In line with this, a holistic sense of nature—in contrast with a scientific way of regarding nature—is put forward by Bonnet (2007), who argues for "a kind of knowing in which personal, moral and aesthetic dimensions are embedded, i.e. a knowledge of things in which 'fact' and 'value' are not separated and things are perceived in their *life*, wholeness, and inherent mystery" (p. 714). He suggests

> a need to re-evaluate the knowledge that we possess through bodily contact with the world. In feeling the resilience of this piece of grass underfoot, this piece of earth to the spade, this piece of wood to the chisel, in feeling the growing chill in the air and apprehending the brooding presence of storm clouds, we engage with the world less through a cognitive ordering and more through a receptive *sensing* that is less susceptible to abstract generalization and objectification. (p. 716)

In line with Dewey, Wickman (2006) claims that aesthetic experiences are involved to a large extent in science teaching and learning, and therefore need to be taken more seriously in science education. He states, "We also need to make further comparisons as to what difference various kinds of aesthetic experiences could make for learning science, and we need examine more closely what difference a teacher can make in science education by noticing and taking more seriously the aesthetic experiences of students" (p. 166). Accordingly the occurrence of aesthetic experiences in science education in compulsory school was studied by Jakobson (2008). She found that the aesthetic experience is an integral part of teaching, and her conclusions were that aesthetics and emotions are significant both for what children learn and how they learn. It is about both positive and negative experiences. If negative experiences are not transformed into positive ones there is a risk that students do not do what was intended and perhaps even avoid similar situations in the future, learn less and eventually may entirely stop being interested in science. Aesthetic experience is thus important with respect to the direction that student learning takes, Jakobson (2008) suggests. Furthermore, Jakobson and Wickman (2008) claim that according to their study, "emotions, values and aesthetic experiences are interrelated in children's learning and also related to cognitive aspects of learning" (p. 64). Similarly, if only the cognitive aspects are addressed during biology lessons, this might lead to "disconnection with nature" (Barker 2007, p. 148). Taking care of living organisms, such as a pea plant in the classroom, for a period of time, as described in this chapter and elsewhere (Nyberg 2004, 2008; Nyberg and Sanders 2014), or closely observing a blueberry plant in the forest, I would say offer a possibility for the teacher to address not only the cognitive aspects of a science lesson. These educational situations also have the "informal" learning qualities which according to Hodson (2003) "can provide the fusion of the cognitive, affective and social that is too often absent in the classroom but is essential to the kind of radical shift in attitudes and values on which socio-political action depends" (p. 664). Hodson also asserts that "There is a need to look at the wider social, political, economic and ethical issues that surround the practice of science" (p. 647). Likewise, Littledyke (2008) argues that cognition and affection need to be explicitly integrated into science education that aims to achieve environmental awareness.

An Empirical Example: 11-Year-Old Students Observing Blueberry Plants

What, then, could a learning situation look like where scientific exploration is integrated with aesthetics and therefore in line with Dewey's definition of experience, implying that these are inseparable (e.g. Dewey 1929)? I would suggest that the learning situation from educational practice below demonstrates this. The excerpt is a transcription from a videotaped discussion between two 11-year-old students (grade 5) during an outdoor excursion in the spring. Students in groups of two or three were asked to choose a place to sit on the ground among blooming blueberry plants (*Vaccinium myrtillus*), not within hearing distance of each other. The task within each group was to discuss how blueberries were formed every year. They were also asked to discuss whether there would be new seedlings in some way and, if so, how if all of the blueberry plants were taken away with their roots one autumn. The task, which they also received handwritten on a piece of paper, was briefly introduced by the researcher, letting the group members taste some frozen raspberries (since no blueberries were at hand) and asking them to try to remember the taste of blue berries. No other instructions were given, apart from checking that everyone had a hand lens. The length of the discussion for each group was around 20 minutes. Three of the discussions were video-recorded and six were audio-recorded. The excursion area was a mixed forest within walking distance of the school. The day was mostly sunny (albeit a little cold) and the place in the forest where the students were seated was quiet, with the vague sound of a road in the distance, but with the most distinct sounds being of birds singing (Fig. 1).

In the following transcribed video-recorded excerpt, two girls start by looking around at the blooming blueberry bushes, and one of them eventually says that she thinks that the new blueberries come from the flowers in one way or another, pointing at something "white". While discussing this they find a "rotten" blueberry, which is also discussed and thoroughly looked at with their hand lenses. Another theory comes up: that the blueberry plants might also spread by roots. However, as the excerpt shows, not only scientific concepts as such are discussed.

Fig. 1 11-year-old students exploring blueberry plants. Photo: Eva Nyberg

A: But I think that, the new bluberries come from something in…
B: Other blueberries…
A: Laughter, and looks up surprised: Other blueberries?
B: Laughing: "New blueberries come from other blueberries…
A: … perhaps from these flowers somehow. I do not know what in the very flower … perhaps is like…
I would have thought that they come from this white little thing sticking out there … then it grows … or something … the white thing furthest in, I think … the one that looks like this (holding hands cupped).
B: Yes.
A: Then perhaps a blueberry is formed there and then perhaps it like grows out from something … out of this white … and then it gets longer and then…
B: Or it spreads by the roots…
A: That could be one way, too…
Did you see where the rotten blueberry disappeared (looking among the shrub)? Did you see it?
B: (Folds back the shrub looking on the ground.) Here…
A: Check it out a bit … (receives it and observes it with a hand lens).
B: But what is this (discovers something in the shrub)? Like white … Have you seen this one?
A: It might be the very beginning of the flower … (observing the rotten blueberry with the hand lens again).
B: (Observing what she has found with the hand lens.)
A: But if you look here in the blueberry there is like a little hole … then it goes down a little like a … (making a descending gesture with her hand) … what can you say … a bit like a … well … a hole.
A: (Giving something to B, presumably the rotten blueberry) Hey, check this one out…
B: (Saying something inaudible)
A: I still think it's cool these kinds of stems, for it's not round and as they normally are, but a little edgy (both of them looking with their individual hand lenses).
A: One can see as well that there will be flowers…
B: There are some cute (?) things on them here, that you almost do not see without this kind of (?). Look at these leaves (handing over a leaf or a twig to A)!

A: Yeah. The leaves are more like shimmering orange-red on one side, then green on the other.
A: Hey .... these are very nice looking actually ... Is this what is becoming the bluebrer ["Blåbret"]?
B: (Mimics) bluebrer ["Blåbret"] ... (laughter)
A: But where ... are they? (looking) ... The skin around the flower. Ifyou look at this, it is quite pretty. It looks kind of like a...
B: Can I have (reaching out her hand and waiting)?
A: It looks like one of these lofty, like a wine glass, round and like ... bananas that go around like this or something (showing with her arms and hands).
B: Like a pineapple...
There is something crawling around inside ... Have you seen?
A: It might be a raindrop ... Maybe it's from where new blueberries grow.
B: From this tiny wine-glass-shaped ... whatever it is...
A: Ey, check this. You ... this rotten bluebrer...
B: (Laughter)
A: BLUEBERRY ... Yes ... it had these holes in it ... Look there between the orange there. I removed this little thing.
B: That one ... humph?
A: Did you see it? The white one ... (goes on exploring—a new flower probably)
B: (Observing with the hand lens) ... Yes...
A: I am sure it's the white thing here.
B: Ah, now (?)
A: Can you see it ... I am sure it is this white thing here that turns into a blueberry and then this ... little flower here ... also becomes blueberry and then it turns into one each (?) ... seems like anyhow ... (?)
A: (Holding up some shrub) ... like bells on ... growing quite closely together (?) ... It is probably rather ... blueberries do not grow that together ... closely...

Here the students explore their hypothesis that the blueberries are formed by the flower (the wine-glass shaped structure), although they do not yet know how pollination and fertilization work. In doing this they seem to discover the stamens ("the orange") and the pistil ("the white thing").

Although the task is strictly scientific—that is, to figure out the reproduction of blueberry plants the students while attending to this, repeatedly notice shapes and colours, and make an effort to describe what they perceive, closely observing the plants. Thus aesthetics seem here to be clearly integrated with the scientific exploration in question. I interviewed the two students three years later, when they were 14, about the spring term when I had been visiting their class on several occasions as they were studying biological lifecycles. During the interview, "A" spontaneously mentioned the situation described above when I asked what she remembered from that spring, and she specifically described the wine-glass shape of the blueberry flower, possibly implying that this had been a special discovery for her.

Aesthetics and Ethics Related to Environmental Awareness

As previously mentioned, the intellect and the emotions, according to Dewey (1916/1999), are often considered to be separated in educational practices: "The intellect is pure light; the emotions are a disturbing heat" (p. 335). He argues that there is no such division and that therefore this, and other views of dualism, should be replaced "by the idea of continuity" (p. 336). Indeed, Manni et al. (2016) conclude in their case study of meaning-making in environmental and sustainability education that they have shown that "emotions and values are relevant and an inseparable part of students' meaning-making processes and a foundation for values" (p. 12). In fact, Dewey's standpoint is that "Emotion is the moving and cementing force" (1934/1987, p. 49) when it comes to experiencing. His notion of emotions as being inseparable from the intellect is also expressed regarding our place in nature: "Fidelity to the nature to which we belong, as parts however weak, demands that we cherish our desires and ideals till we have converted them into intelligence, revised them in terms of the ways and means which nature makes possible" (1929, p. 420).

The significance of human relationships to nature when it comes to environmental awareness is stated by Bonnet (1999), who asserts that "we need to conceive of sustainability not simply as a policy designed to achieve a certain state of affairs, but as an attitude of mind, a way of relating to nature/the environment" (p. 319). In addition, he claims that "nature is that which can never be fully known, intellectually possessed" and that "a proper attitude to nature cannot simply be some kind of extension of a human ethic" (2007, p. 713). Bonnet (2012) argues for the development of an "ethical response" (p. 294) towards the place we live in and an "ethical concern" (p. 295), which comes from our involvement in a place.

He claims that "there is an intimate reciprocity between ethical and environmental concern that fundamentally initiates the character of our caring" (p. 295). Hodson (2003) argues for a biocentric ethic and expresses a concern for both man and other species. He claims that "Adopting such an ethic means having respect for the intrinsic value of all living things, cultivating a sense of compassion and caring towards both human and non-human species, having a concern for maintaining the existence of biological and cultural diversity, and challenging and rejecting all forms of discrimination" (p. 663).

Empathy and care related to environmental ethics are also dealt with by Littledyke (2008), and he makes a case for direct experiences with nature which, according to him, can lead to informed action through "feelings of responsibility and stewardship in protecting living things and the environment" (p. 12). He further argues for direct experiences of nature, which will "enhance a sense of aesthetic appreciation and awe and wonder over the rich biodiversity, beauty and fascinating range of adaptive features of living things" (p. 11). This is exemplified by a variety of approaches to studying plants, including different kinds of sense experience, such as "smell, vision, taste and touch" (p. 12). The contribution of similar sensory experiences when it comes to drawing attention to plants is also reported by Nyberg and Sanders (2014).

On the basis of a number of empirical case studies, Öhman and Sandell (2015) claim that encounters with plants and animals can create moral reactions and spontaneous care for them, and if these encounters are discussed and reflected on there might be implications for young people's development of "environmental morality and awareness of environmental ethics" (p. 259, translated from Swedish). In their empirical study of moral relations in encounters between students and nature, Andersson and Öhman (2015) conclude that their results indicate that "encounters with nature can widen people's environmental and ethical perspectives" (p. 326). Similarly, Barker (2007) holds that "Students need to encounter living organisms first hand if we want them to value living things, nature and the environment" (p. 148), and that this "provides us with opportunities to reconnect young people with nature" (p. 149).

To what extent these experiences really have an impact on future actions such as pro-environmental behaviour, or if environmentally friendly norms attained in one situation become guiding principles in another, we do not yet know (Öhman and Sandell 2015). However, researchers in the field seem to agree that encounters with nature have potential regarding education for sustainability (e.g. Littledyke 2008; Sandell and Öhman 2010; Öhman and Sandell 2015), but that more research is needed to understand when, in what respects and under what circumstances these encounters in and with nature can lead to, for example, long-term positive attitudes to nature and the environment, and to behavioural change (Rickinson et al. 2004; Sandell and Öhman 2010; Öhman and Östman 2015).

An Empirical Example: Student Teachers Observing and Taking Care of Pea Plants

What, then, could a learning situation look like where the learning of scientific concepts are integrated with aesthetics and where ethics also comes to the fore? In my previous research with 11-year-old students (grade 5, observing and taking care of pea plants) for a period of 13 weeks in their classroom, I showed that the entries in their notebooks contained *emotional and aesthetic expressions*, expressions of *caring* and descriptions of the *biological development and physical features* of the pea plants (Nyberg 2008; Nyberg and Sanders 2014). The analysis showed that all three aspects were present in almost all notebooks, and hence that the students had a relationship with their individual pea plant, that they cared about it and that they were interested in its growth and physiological features (Nyberg and Sanders 2014).

Below I will present some excerpts from diary notes from a similar task, but performed by student primary teachers observing and growing pea or bean plants. This is a compulsory assignment within the student teachers course in science education at our university for the teacher programmes grade 1–3 and grade 4–6. The task involves individually sowing a pea or a bean, observing its growth and making (and documenting) observations until the plant flowers and new peas or beans are formed (7–10 weeks). The sowing is done on campus, the observation at home. The students can write their diary notes either electronically or by hand. They are encouraged also to use other forms of documentation, such as photographs and sketches. At least one note per week is required. The task is graded only as a pass or a fail. In spite of this, students often elaborate their diaries and evidently put a lot of effort into the task.

Having read through a number of diaries over the years (around 100 per year) and in this instance 55, it is evident that in every one, expressions of caring and aesthetic experiences, as well as scientific observation, are present. However, depending on the different ways in which the students write and make their notes, all three aspects are not always present in every entry. The selection given below was therefore made with the purpose of finding entries where all of the aforementioned dimensions are visible (Figs. 2 and 3).

Student Teacher 1

Thursday October 15th

It feels like my plant has been looking the same for a great number of days. Today, however, I discovered something that seemed interesting precisely between two leaves. It might be the beginning of something new!

Sunday October 18th

I think, if I am not entirely mistaken, that a flower is on its way! I was beginning to fear that nothing would ever happen, and all of a sudden it does! I have studied the plant and it seems like I have a flower on its way in three different places. To be continued...

Fig. 2 Student teacher's photo of her bean plant with a flower. Photo: Malin Foucard

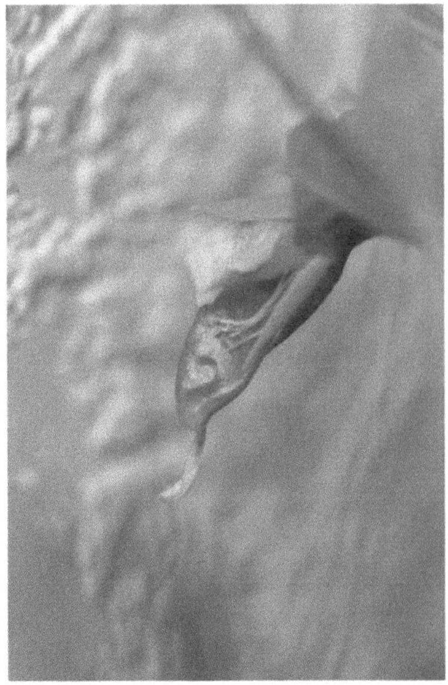

Fig. 3 Student teacher's photo of her pea plant, with a withered flower and the peapod starting to develop. Photo: Elisabeth Altby

Monday October 19th
Now three white pea flowers are beginning to flower! I am so happy!
Wednesday October 21st
This will be the last entry because tomorrow I am handing in my log book. The plant seems to be feeling very fine and when I extend a stalk to measure it, I get it to about 50 cm! The flowers are growing and I look forward to the further development when hopefully there will be peapods. I have enjoyed looking after a pea plant!

Student Teacher 2
Tuesday September 22nd
It is great that the pea plants grow so rapidly! Plant 1 and two have developed new leaves and tendrils and they are now about 28 cm long. I have never been a person who has managed to sustain a plant for more than 2 months and I am a little proud. This is a fantastic way for me to later on in professional life give the students a feeling of pride and importance. The thought that it is me who has made the plants survive is fun!

Student Teacher 3
September 27th Sunday

The pea seedlings have not grown much higher, they are now about 30 cm. However, they have developed a lot of shoots and numerous tendrils stick out from each end of a shoot. The tendrils are delicate and get tangled up. The shoots have also begun to get thicker stems and therefore have become heavier. As a consequence the plants cannot continue to grow upright and have begun to lie down. I'll see if I can build a scaffold for them next week. A fun observation is that the leaves growing opposite each other at the shoots, together look like a butterfly. So it looks like a lot of green butterflies among the shoots.

Student Teacher 4 (Written By Hand and with Clarifying Illustrations)
Day 25 (October 1st)

The pea plant is growing and growing. The leaves that have come out look like butterflies. It is very beautiful and I will always want to have a pea plant in my kitchen window. [An illustration explaining the "butterfly leaves".]

Day 29 (October 5th)

The plants, the two that I have got are just growing and growing. I am absolutely fascinated by the fact that two such beautiful plants can grow out of two little peas. Apart from "butterfly leaves" as I call them, "shoots" or something similar is growing out of the end of some branches. [An illustration explaining this, showing shoots with tendrils.]

Day 36 (October 12th)

The pea plant is growing and growing. It is important to water and sprinkle the plant to keep it moist. Otherwise it will begin to wither. Flower sticks have been very helpful and the plants are winding themselves around the sticks. (Tie to the sticks.) At the tip of each branch further branches arise and these branches get entangled, either with themselves or with others. They also wind around the flower sticks.

These examples illustrate that during this long-term observation (around 7–10 weeks), the learning of scientific concepts related to the growth and the development of the plants seem to be integrated with aesthetic experiences as well as with caring for the plants. There are also entries showing the satisfaction of the students when their caring is successful—that is. when their plants grow well, or flowers and peapods eventually develop. The students' caring is manifested in watering the plant, putting it into sufficient sunlight and various inventive ways of supporting the plant. Whether this caring is so called "natural caring", which, according to Noddings (2013), is a spontaneous action, or "ethical caring", which is the result of a deliberate act, which requires effort to pursue, is not certain. However, in the student teachers' diaries it seems to be the result of a feeling of "I must" take care of this plant and thus could be considered "natural caring" (p. 73).

From a teacher educator's point of view, it is of course satisfying to note that some of the students reflect on the value that this task should have in their future teaching, including the feeling of satisfaction of being able to take care of the plant in a good way. It is also interesting to note the various aesthetic experiences expressed in their entries, as well as the detailed scientific observations, including the questions that these create in the students, and sometimes reflections on the need to understand certain biological concepts to be able to describe what they observe.

Conclusions and Discussion

Having examined some theoretical and empirical research and my own empirical examples in parallel, what can be said about the role that aesthetic experiences have in science teaching and learning, and the significance that these experiences might have for developing environmental ethics?

Dewey's notion of experience, implying that emotion, values and aesthetics are all simultaneously present, is, I now realize, clearly in line with my philosophy of teaching and learning. That aesthetics are present to a large extent in science teaching has previously been shown by, for example, Wickman (2006) and Jakobson (2008). The empirical examples presented in this chapter underline this, and also that, aside from aesthetics, ethical dimensions such as the ethics of caring (Noddings 2013) seem to be present. Interestingly it does indeed seem as if in these situations scientific and aesthetic knowledge do not become "separately and exclusively boxed" as often is the case according to Cobern et al. (1999, pp. 542–543). On the contrary, there are indications that the students and student teachers integrate "the important concepts of their own worlds with the important concepts of science" (p. 557).

The educational situations described here also underline the significance of careful and thoughtful staging to support students' learning and experiencing, as is emphasized by Dewey (e.g. 1938/1997). The important role of the teacher in supporting learning, including the deliberate balance between instruction and experiencing, is also stressed by others (e.g. Piaget 1962; Karplus 1965; Bruner 1985). This was likewise evident in my own case study of teaching and learning in elementary school (Nyberg 2008). The crucial role of the teacher is furthermore emphasized by Garrison et al. (2015) and Andersson and Öhman (2015), who in their respective studies give examples of situations in which the teacher could have focused on ethical issues instead of, or as well as, strictly scientific learning. Accordingly, regarding the potential of the encounters with "nature" in my examples from the educational situations described above, in order to develop any kind of environmental ethics, an explicit discussion would probably be needed. It is thus up to the teacher to bring these issues to the fore when they become apparent in an educational situation. However, there is a possibility that the aesthetic experiences as such in the close observation of the blueberry plant and in the long-term caring of the pea plant could lead to some kind of ethical reflection among the students and student teachers, and in some cases also result in long-term memories of the situations described. This is indicated by the anecdotal evidence presented previously regarding the student who had a strong memory of her observation of the shape of the blueberry flower three years after the educational situation had taken place.

To conclude, it seems that in the examples from the learning situations from educational practice that I have described, science learning and observation are integrated with sensory experiences, such as aesthetics and emotions, and hence have the quality of educative experiences which Dewey argues for. I hope that my examples have shown this, as well as the potential that these learning situations possess regarding environmental ethics. They might thus serve as a possible starting point for framing humans' relationship with nature.

Acknowledgements I am especially grateful to the students and student teachers who have allowed me to use their recordings, diaries, photos and drawings for this chapter. Warmest thanks also go to the editors of this volume as well as the authors, who have pushed me to clarify my positions, especially Dawn Sanders and Marie Grice, who as critical friends have repeatedly given me advice about my future direction. I should also like to thank the Department of Pedagogical, Curricular and Professional Studies at the University of Gothenburg for support of different kinds in writing this chapter.

Note

1. An aesthetic experience is here used in accordance with the definition in Wickman (2006): "a situation as part of an activity where people use aesthetic communicative expressions dealing with beautiful/ugly or pleasure/displeasure" (p. 31).

References

Andersson, K., & Öhman, J. (2015). Moral relations in encounters with nature. *Journal of Adventure Education and Outdoor Learning, 15*(4), 310–329. doi:10.1080/14729679.2015.1035292.

Barker, S. (2007). Reconnecting with nature. *Journal of Biological Education, 41*(2), 147–149.

Bonnet, M. (1999). Education for sustainable development: A coherent philosophy for environmental education? *Cambridge Journal of Education, 29*(3), 313–324.

Bonnet, M. (2007). Environmental education and the issue of nature. *Journal of Curriculum Studies, 39*(6), 707–721.

Bonnet, M. (2012). Environmental concern, moral education and our place in nature. *Journal of Moral Education, 41*(3), 285–300.

Bruner, J. (1985). Vygotsky: A historical and conceptual perspective. In J. V. Wertsch (Ed.), *Culture, communication and cognition: Vygotskian perspectives* (pp. 21–34). Cambridge: Cambridge University Press.

Cobern, W. W., Gibson, A. T., & Underwood, S. A. (1999). Conceptualizations of nature: An interpretive study of 16 ninth graders' everyday thinking. *Journal of Research in Science Teaching, 36*, 541–564.

Dewey, J. (1916/1999). *Democracy and education: An introduction to the philosophy of education* (New ed.). New York: Free Press.

Dewey, J. (1929). *Experience and nature* (Facsimile ed.). London: George Allen & Unwin

Dewey, J. (1934/1987). *The later works, 1925–1953. Vol. 10, 1934: [Art as experience].* Carbondale: Southern Illinois University Press.

Dewey, J. (1938/1997). *Experience and education* (New ed.). New York: Simon & Schuster.

Eger. (1992). Hermeneutics and science education: An introduction. *Science & Education, 1*, 337–348.

Garrison, J., Östman, L., & Håkanson, M. (2015). The creative use of companion values in environmental education and education for sustainable development: Exploring the educative moment. *Environmental Education Research, 21*(2), 183–204. doi:10.1080/13504622.2014.936157.

Hodson, D. (2003). Time for action: Science education for an alternative future. *International Journal of Science Education, 25*(6), 645–670.

Jakobson, B. (2008). *Learning sience through aesthetic experience in elementary school. Aesthetic judgement, metaphor and art.* Doctoral thesis in science education, Department of Education in Mathematics and Science, Stockholm University, Stockholm.

Jakobson, B., & Wickman, P.-O. (2008). The roles of aesthetic experience in elementary school science. *Research in Science Education, 38*(1), 45–65.

Kronlid, D. O., & Öhman, J. (2013). An environmental ethical conceptual framework for research on sustainability and environmental education. *Environmental Education Research, 19*(1), 21–44.

Karplus, R. (1965). *Theoretical background of the science curriculum improvement study.* Berkeley: Lawrence Hall of Science, University of California.

Littledyke, M. (2008). Science education for environmental awareness: Approaches to integrating cognitive and affective domains. *Environmental Education Research, 14*(1), 1–17.

Manni, A., Sporre, K., & Ottander, C. (2016). Emotions and values – A case study of meaning-making in ESE. *Environmental Education Research.* doi:10.1080/13504622.2016.1175549.

Noddings, N. (2013). An ethic of caring. In R. Shafer-Landau (Ed.), *Ethical theory: An anthology* (2nd ed.). New York: Wiley.

Nyberg, E. (2004). Life cycles – For caring and learning about organisms and their environment – A starting point towards an understanding of environmental issues. In P. Wickenberg, H. Axelsson, L. Fritzén, G. Helldén, & J. Öhman (Eds.), *Learning to change our world – Swedish research on education & sustainable development* (pp. 313–328). Studentlitteratur: Lund.

Nyberg, E. (2008). *Om livets kontinuitet. Undervisning och lärande om växters och djurs livscykler – en fallstudie i årskurs 5* [On the continuity of life. Teaching and learning about the life cycles of plants and animals –a case study in year 5] (Göteborg studies in educational sciences, 271) Göteborg: Acta Universitatis Gothoburgensis.

Nyberg, E., & Sanders, D. (2014). Drawing attention to the "green side of life". *Journal of Biological Education, 48*, 1–11. doi:10.1080/00219266.2013.849282.

Piaget, J. (1962). *Comments on Vygotsky's critical remarks concerning the language and thought of the child, and judgment and reasoning in the child.* Vygotsky Internet Archive/Jean Piaget. Retrieved June 25, 2016, from https://www.marxists.org/archive/vygotsky/works/comment/piaget.htm

Rickinson, M., Dillon, J., Teamy, K., Morris, K., Choi, M. Y., Sanders, D., & Benefield, P. (2004). *A review of research on outdoor learning.* Shrewsbury: National Foundation for Educational Research and King's College London.

Rogers, C. (2002). Defining reflection: Another look at John Dewey and reflective thinking. *Teachers College Record, 14*, 842–866.

Sandell, K., & Öhman, J. (2010). Educational potentials of encounters with nature: Reflections from a Swedish outdoor perspective. *Environmental Education Research, 16*(1), 113–132. doi:10.1080/13504620903504065.

Tripp, D. (2012). *Critical incidents in teaching: Developing professional judgement* (Classic ed.). London: Routledge.

Wickman, P.-O. (2006). *Aesthetic experience in science education. Learning and meaning-making as situated talk and action*. Mahwah: Lawrence Erlbaum Associates.

Öhman, J. (2006). *Den etiska tendensen i utbildning för hållbar utveckling: Meningsskapande i ett genomlevandeperspektiv* [The Ethical Tendency in Education for Sustainable Development. A Practical Understanding of Meaning-making]. Diss. Örebro: Örebro universitet.

Öhman, J., & Sandell, K. (2015). Naturmötets betydelse i utbildning för hållbar utveckling [The significance of encounters with nature in education for sustainable development]. In L. Östman (Ed.), *Naturmötespraktiker och miljömoraliskt lärande* [Nature encounter practices and environmental moral learning] (Studia Didactica Upsaliensia 8). Uppsala: Acta Universitatis Upsaliensis. ISBN 978-91-554-8611.2.

Öhman, J., & Östman, L. (2015). Undervisning och lärande i naturmöten [Teaching and learning in encounters with nature]. In L. Östman (Ed.), *Naturmötespraktiker och miljömoraliskt lärande* [Nature encounter practices and environmental moral learning] (Studia Didactica Upsaliensia 8). Uppsala: Acta Universitatis Upsaliensis. ISBN 978-91-554-8611.2.

Östman, L. (Ed.). (2015). *Naturmötespraktiker och miljömoraliskt lärande* [Nature encounter practices and environmental moral learning]. Uppsala: Uppsala Universitet, Acta Universitatis Upsaliensis.

Closing Remarks

Christina Osbeck and Olof Franck

This anthology has, through its various chapters, shown four tracks in ethics and sustainability education. Introduced in the Preface, these together form an approach to and perspective on shaping education and research in this field. The first can be understood as an overarching perspective in line with continental didactical traditions that stress questions about what, how, why and for whom in all contexts where issues about teaching and learning are at stake. The second stresses the focus and content of ongoing learning processes and to what extent they are of analytical or normative character. The third is about how these processes develop and to what degree young people's integrities are being sheltered in these processes. To what extent is there room for them to openly and critically explore different positions and develop standpoints of their own, and to what degree is there an already defined point of view that everyone should develop? The fourth track is about sustainability itself, a phenomenon that is hard to define and that needs to be explored by a hermeneutics of suspicion because different and conflicting interests can be expected to operate behind its taken-for-granted goodness. Critical glasses are necessary for

C. Osbeck • O. Franck (✉)
Department of Pedagogical, Curricular and Professional Studies,
University of Gothenburg, Gothenburg, Sweden

O. Franck, C. Osbeck (eds.), *Ethical Literacies and Education for Sustainable Development*, DOI 10.1007/978-3-319-49010-6_10

the educational practitioner as well as the researcher. In sum, one can say that the four tracks to a large degree concern the overarching didactical questions about "what"—that is, both the meaning of sustainability and analytical vis-à-vis normative focuses—and "how"—that is, the integrity of the individual that is affected by such learning processes.

The chapters have approached these tracks in a variety of ways by foregrounding them at different levels. Beyond these tracks the chapters also show (some more than others) the importance of contextualization in both education and research. Through, for instance, Chap. 3 one realizes that planned and explicit values education has to take the children's ongoing implicit values education into account but also that so called "implicit" education may be very "explicit" from the children's points of view. Research that aims to describe ongoing values and sustainability education could not in order to do so in a full sense be content with describing the explicit processes. This is a conclusion in line with research that in a school context underlines the importance of the hidden curriculum. In this anthology the contextual dimensions are also emphasized in Chap. 6 when stressing the essentiality of working with a life-world perspective where young people's broad experiences are taken into consideration and their way of experiencing the sense of life is stressed. In Chap. 9 we are given examples of what this can mean in practice since it shows through examples from the author's own education how concrete experiences have been of vital importance for students' understanding and development. The meaning of such experience can moreover become especially clear on certain occasions, which is stressed in Chap. 2 through its presentation of the concept *le moment*. How intensive experiences and insights from fiction reading stand in relation to real-life experience is a difficult question, which several contributions address (Chaps. 4, 5, and 8). Is it possible through fiction almost to experience experiences that one has never had and overcome contextual limitations, and in that case what does this mean for sustainability education and for young people's integrity in these processes (see e.g. Chap. 1)?

In the Preface we stress how ethics transcends all of the traditional cornerstones of sustainability (the ecological, the economic, the social and the cultural). Ethics cannot be understood as another cornerstone that can be added to the others. Questions about right and wrong, good and bad, are questions that to a greater or lesser degree are present in every process that people are involved in. Education and teaching are generally a profession that in its very nature is moral, not only in its form but to a large degree in its selection of content. Sustainability is, as this anthology stresses, a complex phenomenon (see e.g. Chap. 7), and what perspectives—what knowledge—that education makes available and thereby possible for the students to gain is of course of importance. It will for instance affect to what degree they are able to realize the complexity and many-sidedness of sustainability and to participate in quality discussions about possible fruitful meanings. An ethical ability is often emphasized as being related to sensitivity and perception, which in turn can be understood as being related to a knowledge and understanding of how things are interconnected.

The habit of raising questions in a conscious sense about right and wrong, good and bad, what characterizes a good human being, a good life and a good society can be cultivated. A school class as a community can collectively develop a sensitivity to meanings and consequences of acts, cancelled acts and available alternatives. In such processes one's imagination, which, for example, fiction helps one to develop, constitutes a great resource. This anthology is a contribution to crucial ongoing conversations about available and sustainable common lives, about how education can make a difference in building prerequisites for such conversations in young people's everyday life-worlds and about how research can contribute to this process. It emphasizes the importance of further research in line with how sustainability education through this volume has come to be understood—that is, as a contextual, complex, normative and analytical as well as transdisciplinary task where ethical considerations are constantly present—not least regarding the centrality of young people's integrities being preserved in such educational processes.

Afterword: Ethical Literacies and Sustainability Education: Young People, Subjectivity and Democratic Participation

Arjen E.J. Wals

We have entered the Anthropocene (Gibson et al. 2015), an era of human-caused global systemic dysfunction where the same species that caused this dysfunction also has the responsibility to turn the tide and respond. How to live lightly, equitably, meaningfully and empathically (i.e. towards the past and the future, towards different cultures, the non-human and more-than-human world) on Earth is the key question of our time. Young people in particular might feel overwhelmed by such a heavy existential question as they have a full life ahead of them and may have serious doubts about having children of their own some day in the face of the declining state of our planet. How can schools help young people to engage meaningfully in such a loaded question? Or, morally speaking, how can they choose not to help them with this question or, worse, make them powerless witnesses and accomplices to this planetary demise by ignoring this question altogether and sticking to "education-as-usual"?

This edited volume brings together authors who are looking for principles, foundations and processes that enable educators, in a broad sense, to connect young people with the key questions of our time. What does

A.E.J. Wals (✉)
Department of Pedagogical, Curricular and Professional Studies, University of Gothenburg, Gothenburg, Sweden

O. Franck, C. Osbeck (eds.), *Ethical Literacies and Education for Sustainable Development*, DOI 10.1007/978-3-319-49010-6

literacy mean in times of fluidity of meaning and the blurring of boundaries? What makes such literacies become ethical or, for that manner, less ethical or even unethical? Is there such a thing as democratic participation in a society that is "rigged" (to use one of the most used words in the recent US elections) towards elitist interests, corrupted by power and fuelled by distrust? These are all difficult, interrelated questions that do not have single or simple answers that can be frozen in time and spread across the globe as universal truths. Yet they are critical questions for educators in general and for sustainability educators working with young people in particular. They are so critical because they provide an entry point into a highly charged "grand dilemma" that seems to be ignored in the emerging field of education for sustainability or for sustainable development for that matter: what is it that needs to be sustained and what is it that needs to be disrupted or transgressed?

This collection written by authors connected to a Swedish international university that seeks to engage meaningfully with sustainable development—the University of Gothenburg—begins to address these difficult questions and by doing so provides a pathway into what Stephen Sterling sometimes refers to as "deep sustainability learning" (Sterling 2008), which addresses ontological questions about our existence as entangled in multiple realities and ways of knowing, and the blinding insights that govern our actions in education and beyond. The first so-called "track" in the book contains contributions that provoke a rethinking of instrumental interpretations of sustainability education that tend to have prescriptive and even moralistic undertones. Such interpretations fail to acknowledge the pluralism of thought and being in the world, and by doing so unwillingly amplify what we might call unsustainability.

In the second track this tension is developed further with a much needed normative probing of how to decide what is moral, right, ethical, fair and just. Authors are reflecting on the role of education in asking these questions and helping learners engage in them. Here I am reminded of a paper I was asked to write as a response to a special issue of *Environmental Education Research* on (environmental) education for sustainability in the Nordic countries—to be more precise, in Denmark and Sweden. The article, entitled "Between knowing what is right and knowing that is it wrong to tell others what is right: On relativism, uncertainty and democracy in E(E)SD", highlights the tension between the realizations on the one hand that "Earth is dying and we must act now" and on the other that "using education to tell people how to live their lives is counter-educational in itself and a possible set-up for indoctrination and inculcation and the loss of democracy" (Wals 2010). The contributions from Sweden and Denmark tended to embrace a Bildung perspective of education (Biesta et al. 2013)—one that is more preoccupied with creating space for learning and democracy than with realizing specific predetermined learning outcomes. This preoccupation can be seen as stemming from having faith in the learners, trust in the teachers and the belief that the freedom to learn with and from each other in spaces that are conducive to exploration will inevitably lead to good outcomes, whatever they might be. However, this preference also seems to be rooted in an aversion to eco-totalitarianism and elitism.

There appears to be a downside to this post-modernist perspective because it implies that "anything goes", that your view is as good as mine, that we must be able to agree to disagree and that respectful dissensus is fine and even desirable. This makes judging problematic. Am I, in searching for a more sustainable world, "an ontology behind" those who have a more relational view of the world and are on a par with the non-human and the more-than-human world? Is a new materialist perspective preferred over a post-human perspective? Are some positions more moral than others, more ethical? And what about ontologies that we have yet to become aware of? This is an enormous challenge for environmental and sustainability educators. When, for instance, do we have the moral authority to say that one ontology is (more) right than others when the goal is moving towards a more sustainable world for all? Should educators not take an explicit stance in this respect and "just" engage learners in conversations, philosophical investigations and biophysical explorations in a way that will "lead" them to a more conscious way of being in the world, leaving it up to them to figure out whether they are comfortable with that or whether they prefer to transform to another way? This is what seems to be the main subject of track three.

Not surprisingly, undoubtedly, the many questions raised so far in this afterword won't be answered here. The point of emancipatory education is not so much providing answers but rather creating dialogue and discursive practices, establishing connections, entangling and untangling, framing and reframing, making the ordinary less ordinary, disrupting the undesirable, seeing and sensing, sensing place and identity, being and becoming, envisioning futures and, indeed, finding moral ground and associated ethical literacies to negotiate what is right in the light of existential questions. This fourth track touches most on envisioning education from such an emancipatory vantage point.

Weaving the tracks together and considering the book as a whole, I should like to draw what we might call a conclusion. This collection represents a plea for an urgent and timely expansion of the notion of literacy by incorporating ethical literacies, not by adding to an already overcrowded curriculum but rather by a systemic reorientation of teaching, education and learning that is less concerned with prescriptive moralistic sustainability education outcomes than with the provision of contexts for learning that afford and invite all of the above characteristics of what I would call emancipatory education with Earth in mind. Such an education is inevitably explorative by nature and needs to be sensitive to the context of which it is part, but what seems to be overarching is the incorporation of philosophical investigations and the reclaiming of intuitive knowing of the world. Paraphrasing Richard Rorty (1998) and linking his thinking to sustainability, we might say that sustainability suffers from our attempts to become more rigorous but benefits from our attempts to become more imaginative. It was, yet again, Albert Einstein who already pointed out that "knowledge is limited, whereas imagination encircles the world" (quoted in Tillmanns 2006, p. 1).

Creating space for imagination and intuitive knowing, combined with what we might call a "planetary consciousness and responsibility that transcends the human", lies at the heart of sustainability education. Such space will need to afford pluralism of ideas and the possibility of ontological encounters. We do not and cannot know what "the best" way of being in the world is, but we do know that the structures and social norms that hegemonic and globalizing Western colonial thought has created over hundreds of years, characterized by hierarchical, reductionist and polarizing ways of thinking, combined with the commodification of virtually everything (water, land, air, bodies, thought etc.), cannot be sustained. A "return to the things themselves" (Husserl 2001) and a reclaiming of intuitive knowing and innate empathy (de Waal 2009) will be necessary. As Martin Buber wrote, "In the beginning is relation – as a category of being, readiness, grasping form, mould for the soul; it is the *a priori* of relation, *the inborn Thou*" (1958, p. 27). Tillmanns believes that it is this *a priori* relation which is the basis for the intuitive knowledge of the world we live in and are immersed in. She argues that we will always retain some form of this intuitive understanding of the world but that it is often replaced by the cognitive skills we develop in school. A consequence of this is that "our cognitive skills are developed in a vacuum, disassociated from our being (2016, p. 3). We might add to this: disassociated from our becoming in a highly politicized world with contested claims and prefabricated moral positions about what is right and wrong.

This book makes a significant contribution to rethinking the role of education in times of systemic global dysfunction. The editors are to be complemented for bringing together such a rich mixture of scholars who walk tracks and pathways less travelled in sustainability education as a contested but undeniably emerging field. Further exploration will be needed both conceptually and practically using the range of innovative methodologies that contributors introduce. Nonetheless, we can already see some kind of convergence towards genius loci-based integral design of schools, urban spaces, homes and workplaces that breathe sustainability, well-being and inclusiveness while recognizing cycles and planetary boundaries. Such a convergence or transition is critical if "we"—all of "us" and all of "it"—are to continue to live on Earth together. Clearly, a systemic reorientation of education will need to take place that will allow young people to find and feel the pulse of sustainability as it expands and contracts with unpredictable rhythms, sometimes harmoniously, sometimes irregularly.

References

Biesta, G. J. J., De Bie, M., & Wildemeersch, D. (Eds.). (2013). *Civic learning, democratic citizenship and the public sphere*. Dordrecht/Boston: Springer.

Buber, M. (1958). *I and thou* (R. G. Smith, Trans.). New York: Charles Scribner's Sons.

de Waal, F. (2009). *The age of empathy: Nature's lessons for a kinder society*. New York: Crown Publishing Group.

Gibson, K., Bird Rose, D., & Fincher, R. (2015). *Manifesto for living in the anthropocene*. Brooklyn: Punctum Books.

Husserl, E. (2001). *Logical investigations* (Vols. 1 and 2) (J. N. Findlay, Trans.). Ed. with translation corrections and with a new Introduction by Dermot Moran. London/New York: Routledge.

Rorty, R. (1998). *Truth and progress: Philosophical papers, volume 3*. Cambridge: Cambridge University Press.

Sterling, S. (2008). Sustainable education – Towards a deep learning response to unsustainability. *Policy & Practice: A Development Education Review, 6*, 63–68.

Tillmanns, M. (2016). *Philosophy for children and their intuitive knowledge Philosophy Now*, 119. Accessible via: https://philosophynow.org/issues/106/The_Need_To_Move_Beyond_Homo_Faber

Wals, A. E. J. (2010). Between knowing what is right and knowing that is it wrong to tell others what is right: On relativism, uncertainty and democracy in E(E)SD. *Environmental Education Research, 16*(1), 143–151.

O. Franck, C. Osbeck (eds.), *Ethical Literacies and Education for Sustainable Development*, DOI 10.1007/978-3-319-49010-6

Index[1]

[1] Note: Page numbers followed by "n" denote notes.

O. Franck, C. Osbeck (eds.), *Ethical Literacies and Education for Sustainable Development*, DOI 10.1007/978-3-319-49010-6